SpringerBriefs in Ethics

Springer Briefs in Ethics envisions a series of short publications in areas such as business ethics, bioethics, science and engineering ethics, food and agricultural ethics, environmental ethics, human rights and the like. The intention is to present concise summaries of cutting-edge research and practical applications across a wide spectrum.

Springer Briefs in Ethics are seen as complementing monographs and journal articles with compact volumes of 50 to 125 pages, covering a wide range of content from professional to academic. Typical topics might include:

- Timely reports on state-of-the art analytical techniques
- A bridge between new research results, as published in journal articles, and a contextual literature review
- A snapshot of a hot or emerging topic
- In-depth case studies or clinical examples
- Presentations of core concepts that students must understand in order to make independent contributions

More information about this series at http://www.springer.com/series10184

Vojin Rakić

How to Enhance Morality

Springer

Vojin Rakić
Center for the Study of Bioethics
University of Belgrade
Belgrade, Serbia

ISSN 2211-8101 ISSN 2211-811X (electronic)
SpringerBriefs in Ethics
ISBN 978-3-030-72707-9 ISBN 978-3-030-72708-6 (eBook)
https://doi.org/10.1007/978-3-030-72708-6

© The Author(s), under exclusive license to Springer Nature Switzerland AG 2021
This work is subject to copyright. All rights are solely and exclusively licensed by the Publisher, whether the whole or part of the material is concerned, specifically the rights of translation, reprinting, reuse of illustrations, recitation, broadcasting, reproduction on microfilms or in any other physical way, and transmission or information storage and retrieval, electronic adaptation, computer software, or by similar or dissimilar methodology now known or hereafter developed.
The use of general descriptive names, registered names, trademarks, service marks, etc. in this publication does not imply, even in the absence of a specific statement, that such names are exempt from the relevant protective laws and regulations and therefore free for general use.
The publisher, the authors and the editors are safe to assume that the advice and information in this book are believed to be true and accurate at the date of publication. Neither the publisher nor the authors or the editors give a warranty, expressed or implied, with respect to the material contained herein or for any errors or omissions that may have been made. The publisher remains neutral with regard to jurisdictional claims in published maps and institutional affiliations.

This Springer imprint is published by the registered company Springer Nature Switzerland AG
The registered company address is: Gewerbestrasse 11, 6330 Cham, Switzerland

"Rakić explains with clarity and rigour how moral bioenhancement can be motivated even without concern for human extinction and why it can be implemented voluntarily. This book will surely become a major reference for the further discourse on how to become better at being good."
—Anders Sandberg, Oxford University

To Sara, Tea, Filip and Andrija

Acknowledgements

I wish to thank various scholars and friends for the invaluable discussions we had about numerous arguments raised in this book and for the support they gave to its publication. They include, but are certainly not limited to, John Harris, Ingmar Persson, Julian Savulescu, Nicholas Agar, Arthur Caplan, Robert Sparrow, Peter Singer, Anders Sandberg, Milan Ćirković, Katrien Devolder, Amnon Carmi, Harris Wiseman, Thomas Douglas, Bert Gordijn, Thomasine Kushner, Erik Parens, Oliver Feeney, Sarah Chan, Josephine Johnston, Aleksandar Damjanović, Yves Agid, Nada Gligorov, Shai Linn and Russell D'Souza.

For their administrative and logistical support I am primarily indebted to Milica Milašinović, Ana Stanković, Dejan Pejović, Danilo Polić, and Stefan Mićić.

My gratitude goes to Floor Oosting for her superb handling of the publication process, as well as to Christopher Wilby for his important role in it.

I am extremely grateful to my family for their patience and tolerance they had for me during the many years of my devotion to the themes I address in this book. Without them I would not have been able to write this book and the many articles that preceded it.

Introduction

> *The Earth is evil*
> *....Justine's justification of her wish that our planet be obliterated*[1]

The Theme

Human history is marked by a strong interest in moral enhancement. This book will largely deal with the following two aspects of moral enhancement:
1. What does it mean to be good?
2. How to become better?

Aspect 2 will be mostly covered through the lens of biomedically based moral enhancement (moral bioenhancement—MBE).

Humanity's Big Leap

King Shahryar of Persia and his brother Shahzaman were anguished after becoming convinced that their wives were unfaithful. Shahryar decided to adopt a hands-on attitude. Not only that his wife and lover had to pay with their lives, but at the same time, he decided to punish women in general. He became obsessed with a pathological hatred toward the female gender. This led him to spend every subsequent night with a different virgin who was escorted to him by his vizier while ordering at dawn the beheading of each of them. Each night he performed this bizarre ritual. Several thousands of women died accordingly. As the King was running out of virgins, the vizier's daughter Sheherezade came up with a plan. She decided voluntarily to take the huge risk of spending a night with Shahryar, determined to change his cruel heart. Her father (the vizier) objected vehemently, but after Sheherezade's persistent demands, he finally gave in. With the help of her sister, Dunaizad, Sheherezade

---

[1] Dialogue from Lars Von Trier's film "Melancholia".

seduced Shahryar. Already the first night Sheherezade began telling the vile sociopath a story that she did not finish by dawn. Shahryar became interested in how the narrative will end and ordered Sheherezade to finish her not yet completed story the next night. She did that but started that same night telling Shahryar another story only to leave that one also unfinished by dawn. As a consequence, Shahryar found himself delaying her execution each night. Sheherezade's stories covered a variety of themes, many of which had a strong impact on Shahryar's imagination. Shahryar fell in love with Sheherezade. In the end, his cruelty and hatred toward women evolved into something nobler. Shahryar became a better man. The 1001 nights Sheherezade spent with him did the job. They lead to Shahryar's moral enhancement.

Why did Sheherezade do this, why did she risk her life? To save other women? To minimize the number of dead people? To spare her father the King's wrath because he was running out of virgins? To perform her duty? Supererogation? Or was Sheherezade all the time in love with Shahryar? At which point came Shahryar's love for Sheherezade in as a variable that could explain his change of heart? What kind of love, if any, motivated Sheherezade to undertake an enterprise that looked almost suicidal? Could moral bioenhancement, if it were available at that time, have had a similar impact on Shahryar? Could it have made him a better man?

Moral enhancement has been a gargantuan fascination of humanity throughout its history. One cannot remain unimpressed by the amount of literature it has produced on this subject. The aspiration to become a better person appears to have been one of the greatest preoccupations and passions in almost all cultures. Not surprisingly, a lot has been written on the subject of how to morally enhance ourselves—the Bible being one of the most influential books on moral enhancement in human history.

A similar question, "how to be good?"[2] (almost a reformulated version of the question of what goodness is), has possibly received less attention than the question of how to *become better*. The reason might be that most people apparently believe to know how to be good, but do not think they act in accordance with that knowledge. Hence, they ask the question how to become better, that is, how to bring into line their actual behavior with their knowledge of what it means to be good.

In spite of all its efforts, humanity has largely failed in morally enhancing itself. Some successes have been booked: slavery is considered nowadays as morally abhorrent (and is illegal anywhere in the world), the number of liberal states has been on the increase in the last two centuries,[3] the history of humanity apparently shows a steady decrease in violence.[4] These successes are however meager ones for a time span of several millennia, a span after which the world still sees horrendous injustices, cruelties, and suffering. An essential (but not only) reason for this scanty record is that humanity has proven incapable of bridging the "comprehension-motivation gap": the gap between how we act and how we believe is morally right to act, the gap between knowing the good and acting good.

---

[2] E.g., the title of a book published by Harris (2016).
[3] See Doyle (1983).
[4] See Pinker (2011).

The sole restriction God gave to Adam and Eve was not to eat from the Tree of Knowledge of Good and Evil. After violating God's only command, Adam and Eve both learned about evil and experienced evil. Knowing the bad and being bad coincided in their case. Before eating from the tree, they didn't know what evil was. They only knew goodness and they were good. At that stage knowing the good and being good coincided. Their sin had therefore, among else, the following consequences: they learned about evil, they became capable of being evil and they developed a gap between knowing the good and being good. In its Biblical interpretation, Adam and Eve were the ones who created the comprehension-motivation gap in morality.

Can new biotechnologies be of help in dealing with this aspect of the Original Sin of humans? Can they help humans bridge what might well be the greatest predicament of their moral existence: the gap between what they do and what they believe they *ought* to do? Can they open up the possibility of humanity becoming better in ways it has not been capable of until now?

Moral bioenhancement (MBE) proponents believe that MBE can make us better, that we will make a leap humanity has never made in its entire history in a realm that is so dear to it. But can humanity do this? Can MBE technologies help humanity make a jump in its evolution it has never made before? Are we about to witness a new humankind consisting of people who act in line with what they know is right? Are we about to become better people? Can we become better people in the short run?

This book will answer the question what humans should do in order to become better. It will examine how to motivate people to become better. The argument will be defended that behavior we consider as moral, happiness, and self-interest operate in a circularly supportive fashion. Morally apposite actions make most people happy most of the time. They are therefore in their self-interest. The human inability to genuinely comprehend this relationship and to act upon this comprehension is one of the greatest mysteries of human existence. This book will show how humanity can make the greatest possible leap in its moral functioning by overcoming this inability.

It will be argued that existing, but mostly future MBE technologies can be of help in this endeavor. Still, they are far from being sufficient to morally enhance humans. They can give humans an impetus to make the big leap. This impetus is essential, but not enough. It will be shown what kind of mechanism is needed in order to create the opportunity for a major moral enhancement of humankind.

The Origin of Morality

An essential question is why humans think that they ought to behave in a way they consider as "moral". Related to this issue is the question what the origin of morality is. The answer can be given on the basis of three core perspectives. At this point, I will only briefly refer to them.

First, the socio-cultural explanation states that morality is a social construct. Indeed, part of our moral functioning develops through socialization. But socialization does not explain enough. There are certain moral rules that apply trans-culturally and that have applied trans-historically.[5] Moreover, the question is whether the social construct of "morality" is not merely convention rather than morality. Convention is a social construct. If we posit that morality is also merely a social construct, we equalize it with convention. Universal (trans-geographic and trans-historical) moral values would in that case merely be frequently accepted conventions, conventions that can rather easily change. This begs the question why some of them apparently haven't changed until now and why they apply irrespective of culture.

Second, a biology-based (evolutionary) perspective argues that our drive to survive has resulted in our empathy extending to our kin. Kin can help us survive and hence they are useful. At a certain point, the argument goes, humans noticed that they do not treat all people equally, that this is wrong, and subsequently, their empathy came to be corrected (enriched) by moral reflection—extending moral rules to an ever expanding circle of humans (for this argument instructive is Persson and Savulescu 2017). The downside of this perspective is that it assumes that humans had a problem with inequality. But nature has created humans unequal. Hence, there is no reason to assume that egalitarianism is a *natural* human inclination. It might very well be a socio-political construct.

Third, the religious perspective starts with the assumption that the criterion by which our moral inclinations, are being assessed is different from these inclinations themselves. One way to explain how such a criterion developed, that is how we obtained it, is to assume that God has given it to humans. The religious perspective assumes precisely that. For example, sometimes we feel the urge or moral inclination to help a drowning person, at the same time being afraid to risk our own life by trying to save her. Our inclination to help a member of our species conflicts with our inclination to preserve our own life. But apart from these two inclinations there is a criterion at work that tells us that it is morally right to help the drowning person (provided, of course, that we have a good chance not to drown ourselves). That criterion is about the *ought*, about what is morally right. The religious perspective maintains that God has provided us with this criterion.

All three perspectives make assumptions. It has been briefly noted what some of the weaknesses are of the first two perspectives. But the assumption of the third can also be disputed. One way to do that is to invoke the issue of theodicee: if God's power and goodness are infinite, why has He allowed suffering of the innocent? We don't think that the innocent should suffer. Does God have a diverse opinion on this? Does that imply that God has different moral rules than humans have? And if we accept the religious perspective, aren't we inconsistent if we think that our moral rules are and should be distinct from God's? To formulate it even more radically, aren't we abandoning the religious perspective if we assume that God provided us with *ought* criteria He doesn't accept Himself?

---

[5]Instructive in that regard is C.S. Lewis's listing in an Appendix to his *Abolition of Man* of a significant number of universal trans-geographical and trans-historical moral values (Lewis 1943).

Indeed, it is difficult to find a cogent explanation for the origin of morality. Explaining the origin of morality requires the understanding of a complex system. It is questionable whether we can explain a complex system of such a magnitude and with such qualities. What remains however is the possibility to discard morality altogether or to legitimize it. Discarding morality altogether does not seem to be something the majority of people can live with. Legitimizing morality can be done by understanding its principles and by applying them. Hence, if we want to legitimize morality, we have to learn what it means to be good *and* to be good.

It can be argued that the origin of morality resides on love. To desire things to happen to others that one desires to happen to herself is a principle that is an essential feature of the concept of love. The wish that justice be done to those who made innocent people suffer is based in love for those people, and perhaps for humans in general. As love is an important human inclination, and as morality is intertwined with love, it is difficult to discard morality.

Moreover, both morality and love are acts of our will. Love for one's children, spouse, parents, siblings, friends, or nation are types of love that can be called *sentimental*. They are based on our emotions. Love for all people is however based on a decision to love all humans. In that sense, it is not sentimental, but rather *volitional love*. Morality operates in a similar fashion. Morality that extends to children, spouse, parents, siblings, friends, and nation, we can call *sentimental morality*. Morality extending to all we can call *volitional morality*.

Additionally, we can stimulate volitional love by behaving *as if* we love. After some time our kind and loving behavior might be transformed into genuine love. That links love and morality as volitional acts from another angle: by acting *as if* we love we will not only become truly more loving but as a consequence of that, also more moral. Two essential questions are why so many humans do not behave *as if* they love and whether that will change in the future, enhancing both their volitional love and their morality. Chapter 2 will deal more extensively with this issue.

Kant

One of the great ethicists who believed that humanity will become better in the future was Immanuel Kant. His conception that best describes a human community of the future is that of the *Ethical Commonwealth*. It is a commonwealth of people who *act* "united under laws without being coerced, i.e. under *laws of virtue* alone" (Kant 1907, Ak. 6:95). It is a community of people who have succeeded to morally enhance themselves by bringing their behavior in line with what they know is right. They have transcended the discrepancy between knowing the good and acting good. They have surpassed the comprehension-motivation gap.

Kant's expectation of a better world based on humanity's moral progress rested on an argument that was rather similar to his arguments in favor of the existence of God and the immortality of the soul: we have a moral duty to believe in these conceptions and that is allegedly the reason why these conceptions are true. Such an argument can better prove the future of the Ethical Commonwealth than the existence of God and an immortal soul in us: belief in a morally just world order might induce us to realize it, while mere beliefs in God and an immortal soul will not turn them into reality.

Nowadays, however, there might be more effective ways than mere belief is in order to realize something akin to Kant's Ethical Commonwealth. *In this book, it will be argued that new biotechnologies can forcefully boost Kant's line of reasoning and his belief in moral progress resulting in a new moral order in the future.* Moral enhancement by biomedical means will have the opportunity to take the role that belief (in a morally just world) had in Kant's thought.

New Biotechnologies for the Enhancement of Human Morality: The Current Debate

A lot of new biotechnologies open up new moral challenges but offer also new moral opportunities. The new challenges and opportunities are numerous and extend to various technologies. Some of them are in the realm of MBE technologies.

Taking a closer look at the moral challenges of new biotechnologies, we see that they are expanding at a fast pace. In 2004, Ed Silverman published an article in *Biotechnology Healthcare* called "The 5 most pressing issues in biotech medicine" (Silverman 2004). The 5 issues Silverman put forward were protecting human subjects in clinical trials, affordability (of biotech medicine), privacy, stem cell research, and defending the United States against bioterrorism. Although these issues remain pressing, they have acquired stiff competition from numerous other pressing issues. They include new genetic engineering technologies, ever more sophisticated cognitive enhancement technologies, new tools in reproductive medicine, genome editing for therapeutic and enhancement purposes, issues of who owns our genetic data, pandemics not related to bioterrorism, MBE technologies.

The possibility of novel MBE technologies is one of the issues that have attracted highly intense and controversial attention in recent years. These technologies belong to those that pose various moral challenges, at the same time offering opportunities for solving them. They offer new opportunities to humans to become better.

In what follows I will expand on three main theories I take issue with throughout the book: the theories of Persson and Savulescu, John Harris, and Harris Wiseman. I selected those theories because they are, first, influential in scholarly literature, and second, they defend quite diverse viewpoints—all of them differing from my position as well. In the course of discussing the mentioned and some other theories, I will hint at my own position, which I will expand on later in the book.

Persson and Savulescu were essential proponents of MBE in recent years. They belong to a group of bioliberal scholars who have adopted a stance favoring human enhancement. MBE belongs in their view to various other enhancement opportunities that are to be pursued. In that regard, Persson and Savulescu are no different from John Harris, by many considered as the pivotal personality of the ethics of enhancement.

They do differ from each other however in that John Harris believes that cognitive enhancement is sufficient to make us better. According to Harris, when we dispose of our prejudices, incompetencies, and "idiocies" (Harris 2011: 108–111) we will become better people. Once we understand, at a cognitive level, that racial biases are immoral, that xenophobia and most other forms of heterophobia are morally wrong and dumb, we will enhance our morality. Hence, cognitive enhancement is the solution to our moral inadequacies. Harris maintains that such enhancement can be both traditional (e.g., education) or biomedical in nature.

Persson and Savulescu see things differently. They diagnose a misfit between the challenges posed by existential harms humanity is exposed to and its moral aptitude to deal with them. Humans have been adapted in their evolution to extend their moral perspective to those who are near and dear to them and to the relatively close future. Existential harms posed by new technologies require however a different sort of morality, one that extends to humanity as a whole and to the farther future. As new technologies develop faster than our morality, the latter's enhancement ought to be accelerated. The best way to do that is through the application of new biotechnologies. Oxytocin, SSRIs, dopamine, as well as other substances and technologies that can increase our empathy and consequently strengthen our altruism, while subduing aggression, are to be made mandatory (Persson and Savulescu 2008).[6] In that way, we will lower the likelihood of what Persson and Savulescu call "ultimate harm".[7]

In Savulescu and Persson (2012) even the design and implantation in human minds of a "god machine" is being advocated. This device controls our thoughts and as soon as it discovers that we have developed a "grossly immoral" thought, it disables our will to act upon it. In that way, ultimate harm is to be prevented. Any thought that might be harmful enough to cause it, will be policed by the "god machine" and if assessed as sufficiently dangerous for the existence of human (and non-human?) life, the will of the person who has developed it will be impaired to the degree that the "grossly immoral" thought disappears and thus ceases to contribute to the jeopardy of worthwhile life on our planet. As compulsory MBE and the "god machine" are unlikely to function well in a liberal social context, Persson and Savulescu voice their reservations against contemporary liberal societies (Persson and Savulescu 2011).

Since the "god machine" makes people become ignorant about the "grossly immoral" thoughts they have possibly developed, it appears to have the capability of returning humans to the state before the Original Sin. But there are two major differences between that state and the effects of the work of the "god machine". In the case of the "god machine", humans are not free to decide about what they will do. Moreover, the "god machine" makes humans *a posteriori* ignorant about (major)

---

[6]In their later publications, Persson and Savulescu do not take a decisive stance on whether moral bioenhancement ought to be elective or compulsory (e.g., Persson and Savulescu 2011). Nonetheless, they do not abandon the concept of compulsory moral bioenhancement.

[7]Persson and Savulescu define "ultimate harm" as an event or series of events that will annihilate life or make worthwhile life on this planet forever impossible (Persson and Savulescu 2014).

evil—by "deleting" the grossly immoral thoughts they develop. In the Garden of Eden, on the other hand, the human was free, but a priori ignorant about evil. He (and she) did not know about evil. We will turn to the issue of the relationship between God (Christian, Jewish, Islamic) and the "god machine" later in this book.

The moral theories of both John Harris and Persson & Savulescu offer important insights and possible solutions to a number of ethical issues, but at the same time, they face serious difficulties. John Harris has devoted a lot of his attention to prejudices and conservative illusions impairing the implementation of scientific discoveries that benefit humanity. His arguments dealt with the ethics of human enhancement, in-vitro fertilization, MRT, CRISPR Cas9, and other technologies in the fields of genomics, genetic technologies, considerable life extension—just to mention a few of them. For decades John Harris has been attempting to show that we have a moral duty to apply a variety of those technologies. And once we divest ourselves of our prejudices and "idiocies" related to science, race or gender, we will enhance our morality. This is an essential point with far-reaching implications in the fields of science and ethics. John Harris ought to be credited for insisting on it.

On the other hand, John Harris failed to acknowledge that prejudices and dumbness at a cognitive level are not the only reasons for our moral inadequacies. He didn't recognize that we frequently do understand what is morally right but that we do not act in line with that understanding. The importance of the comprehension-motivation gap has largely escaped his attention.

The position of Persson and Savulescu can however also be challenged from various angles. First, their initial idea to make MBE compulsory deprives humans of an essential element of their humanity (and of moral behavior): the element we use to call "free will" or "freedom of the will".[8] They wish to use MBE in order to lower the likelihood of humans inflicting ultimate harm upon themselves, but by making it compulsory and thereby depriving humans of their free will they already inflict a degree of ultimate harm upon them. They throw out the baby with the bathwater!

Second, the conception of compulsory MBE is in danger of confusing what is moral with what is legal. Being moral implies both understanding what is moral and acting in line with this understanding. As has been noted already, knowing the good without acting good is not sufficient for being moral. Conversely, being induced to act good without knowing the good is also not sufficient for moral enhancement. Being coerced to be moral is not genuine moral enhancement, since it implies that an external mechanism (one that is not ourselves) subjects us to moral behavior. If we don't want to subject ourselves to such coercion, we will suffer certain sanctions. That is precisely what a legal system does. A legal system cares about our actions, not about our motivation as to why we act in line with the law. In that sense, it is similar to the "god machine". That is however not how morality works. For morality, it is not only actions that count. Our motivation is important as well.

Third, the "god machine", a device that is an extension of the conception of compulsory MBE polices and "deletes" our thoughts and hence does not only impact on what we will but also on what we think. It infringes upon our freedom of thought.

---

[8] Further in this book, I will use these terms without quotation marks.

Such device resembles more a police machine than God from the Judeo-Christian and Islamic traditions. This God leaves intact our "freedom to fall". The "god machine", on the other hand, controls our thoughts and changes them if they are judged (by the device) to be "grossly immoral". Hence, it disables us to "fall".

Fourth, a relevant issue is who controls the "god machine". It is unlikely to be the society's "moral elite". Such an elite, even if we were able to locate it, is not the elite with most political or financial power in society. Hence, the device is unlikely to be controlled by the morally most apposite people in society, but by those who already are the most powerful in society. And that is not necessarily the moral elite.

In spite of all the difficulties the designers of the "god machine" face, they cannot do without it: compulsory MBE can work in an efficient and precise manner only with the "god machine". Without a device policing our thoughts all of us would be indiscriminately subjected to the same type of compulsory MBE—something that is morally unsustainable.

Fifth, the conception of compulsory MBE and of a device that polices our thoughts returns authoritarian and repressive practices to our societies. Such practices have been surpassed in most of the developed world. Even if compulsory MBE were necessary to lower the likelihood of ultimate harm, it ought nonetheless to be stated as a matter of fact that it would introduce in our lives something we have believed and hoped is historically behind us.

Sixth, achieving the end of moral enhancement by coercion would radicalize the relation between moral ends and means to an extent that might render the very idea of moral enhancement absurd. If MBE were made compulsory, we would accept not only that it is possible to have an authority that decides about which moral ends are praiseworthy and which means are most effective to attain them, but that it can impose its decisions by coercion. We have witnessed throughout history various examples of ideologies that claim to know which objectives are morally right, while their proponents frequently used coercive means to enforce them. The use of such means that are directed toward moral enhancement runs contrary to the right of humans to freely take decisions on morally relevant issues. Should it be accepted, such an endeavor might utterly compromise the very idea of moral enhancement.[9]

Harris Wiseman is a scholar whose position is on the opposite pole of that of Persson and Savulescu. Wiseman opposes MBE a priori. In his *Myth of the Moral Brain* (2016) he objects to the position of Persson and Savulescu from, among else, the point of view of what ought to be the grounding rationale of MBE. According to Wiseman, that should not be the lowering of the likelihood of ultimate harm. Moreover, Wiseman is also unsympathetic to all those who think that MBE can achieve anything truly meaningful. He focuses on what currently available MBE technologies can or cannot bring about, failing to take into account what they might offer in the future. In fact, he believes that MBE technologies have reached their zenith already.

Wiseman's stance begs the question: why would moral bio-enhancement technologies be an exception to practically all other bio-technologies about which we

---

[9]For this argument see Rakić (2017).

don't believe that they have reached their peak? The fact that part of the difficulty of our existence as moral beings does not reside in our lack of motivation to act morally, but in our comprehension of what is morally right, does not mean that MBE technologies cannot be of help in motivating us to act morally when we know what is morally right—or that they cannot offer more in the time to come.

The shortcomings of the positions of John Harris, Persson & Savulescu and Harris Wiseman open up the space for the idea of *voluntary moral bioenhancement*—a conception that, as I argue for some years already, is the best option we have in order to become better. John Harris thinks that cognitive enhancement is sufficient for our moral betterment, Harris Wiseman doesn't believe in the potentials of MBE, while Persson and Savulescu, using ultimate harm prevention as the grounding rationale for MBE, advocate in one form or another compulsory MBE.

Each of them apparently propose something with at least one serious shortcoming: cognitive enhancement is not enough to become better (it might even be argued that the central trait of the history of morality refers to humans trying to address the predicament of the gap between knowing the good and acting good), compelling people to behave in one way or another encroaches upon their freedom of the will and even their freedom of thought while reflecting on MBE solely through the lens of what currently existing MBE technologies have on offer is a viewpoint on moral enhancement that does not take into account what appears likely to become possible in the not too distant future.

Voluntary moral bioenhancement (VMBE) impacts our motivations and makes us therefore better in a way that is not limited to the enhancement of our cognitive abilities, it leaves open the space to humans to become better without compelling them to do so, and it is oriented towards the utilization of new means of MBE—provided, of course, that they are efficient and safe. Nonetheless, VMBE has its shortcomings. This book will address them, but also show why VMBE is still good enough to opt for it.

The essentials of the conception of VMBE are the following:

1. The notion of VMBE can justify MBE without using ultimate harm prevention as its key rationale. It will be shown why that is so. That is one of the main differences between this conception and the viewpoint of Persson and Savulescu.
2. The conception of VMBE differs from the one that John Harris advocates in that the former does not accept that cognitive enhancement is sufficient for moral enhancement. Cognitive enhancement can help people divest themselves of various cognitive incompetencies that lead to immoral behavior, but by no means does this necessarily lead to morally enhanced *behavior*. The gap between what we do and what we believe we *ought* to do would remain—even if we assumed to have perfect cognitive competencies. Efficient MBE technologies would however motivate us to behave as we think is morally right. Hence, they would help us close this gap.
3. The notion of VMBE focuses not primarily on the effectiveness and safety of currently existing means of MBE, but rather on MBE *in principle*. If the effectiveness and safety of existing MBE technologies are currently questionable,

that does not mean that they will remain so. On the contrary, we are likely to witness their development in the future.

4. VMBE keeps our freedom (of the will) intact, while compulsory MBE does not. In general, MBE might lower the likelihood of ultimate harm in the long run. It might not do so decisively, but again, we have to give up on the unachievable aim of safeguarding certainty of survival of our species. If the cost of this survival is to give up on our free will as we perceive it (as is the case with compulsory MBE), the cost is too high. Humans whose (perception of) free will has been violated have already been deprived of a degree of their humanness. They have faced a form of ultimate harm already.

5. VMBE does not make moral reflection practically superfluous. Compulsory MBE, on the other hand, is not accompanied by appropriate and usable moral reflection, as it deals with individuals who are compelled to subject themselves to it. The VMBE position claims that the main beneficial outcome of compulsory MBE (more safety, according to Persson and Savulescu) does *not* trump its detrimental outcomes: our moral reflection being rendered practically superfluous, in addition to our freedom being diminished.[10]

The conception of VMBE differs from the one advocated by Persson and Savulescu in (1), (4), and (5), from the standpoint of John Harris in (2) and from Wiseman's position in (3). This does not mean that there are no other differences, but the above-stipulated ones are essential.

A theory of VMBE faces however one serious problem. It has to give a persuasive answer to the question what will motivate humans to voluntarily subject themselves to this type of enhancement. If it is not the prevention of ultimate harm, what should be the grounding rationale of MBE?

An extrinsic motivation other than ultimate harm prevention can consist of the state adopting policies that incentivize MBE. Those who undergo MBE could obtain "advantage of opportunity". This can consist of tax reductions, retirement benefits, schooling allowances for their children, housing allowances, as well as various other forms of "positive discrimination" (Rakić 2014: 250). Such policies have however considerable shortcomings. First, it is difficult to imagine that morally unenhanced political decision-makers will adopt morally wise policies. Second, incentivizing MBE implies that we have to put a price on morality—something that appears difficult. Third, a state incentivized program of MBE contains certain elements of coercion. If those who do not undergo MBE are not entitled to the benefits the morally bioenhanced have, state incentivized MBE policies may appear controversial with regard to respect for citizens' rights and freedoms.

This brings us to the seventh essential of a theory of VMBE.

7. Humans can have an alternative, intrinsic motivation to voluntarily opt for MBE. This motivation is based on the positive correlation between acts of goodness and happiness. When we are good, we generally tend to feel happy. Hence, it is happiness

---

[10]In the course of the book, two additional potential detrimental outcomes of compulsory MBE will be added and explained. One is that compulsory MBE might limit the capacity of humans to truly *love*, the other is that it might bring into question *human identity*.

that can be the grounding rationale for VMBE. Goodness, namely, tends to make most people happy, most of the time. Morally apposite lives have the tendency to contribute to human happiness. Conversely, happy people are more likely to be good. Hence, it appears that the relationship between goodness and happiness is circularly supportive. This relationship will be substantiated in Chap. 6 on the basis of various findings that have demonstrated its existence.

Chapter 1 will discuss a variety of arguments that have been proposed in favor of or against enhancement technologies. It will show why performance enhancement is generally not morally controversial. In Chap. 2 the focus will be on moral (bio-)enhancement and morality. The issue will be addressed of what it is that makes us less good than we *can* be and what we may do in order to become better than we are. The role of love in morality will be taken up. The notion of "goodness" will be specified, as well as the concept of the comprehension-motivation gap. The issue of MBE will be addressed in the framework of its relationship to freedom. In Chap. 3 the stances of a number of key proponents and key opponents of MBE will be discussed. They will be divided into four groups: first, those who categorically support MBE; second, those who hypothetically support MBE (i.e., in the context of certain assumptions); third, those who hypothetically oppose MBE; fourth, those who categorically oppose MBE. The conception of Voluntary Moral (Bio-)Enhancement will be introduced. It will be argued that MBE ought to be pursued, provided that it is voluntary. In Chap. 4 the harshest form of MBE skepticism, that is, categorical opposition to MBE will be elaborated and criticized in more detail. The focus will be on the writings of Harris Wiseman and my arguments against his position. Chapter 5 will address the issue of which types of MBE technologies are currently available but also which are realistic to be effective in the future. It will be shown why compulsory MBE is ineffective *in principle*. Chapter 6 expands on VMBE. It will address the issue of what will motivate humans to subject themselves to MBE. It will be argued that ultimate harm prevention is an extrinsic motivation that is less viable than intrinsic motivations for being good. The positive correlation between being good and being happy will be used as an argument for proposing happiness, instead of ultimate harm prevention, as the grounding rationale for MBE. Chapter 6 will be followed by an Addendum in which it will be argued that, apart from VMBE and CMBE, the possibility can be raised of a third type of MBE: Involuntary Moral Bioenhancement (IMBE). IMBE will denote a variant of moral enhancement of the unborn that is neither voluntary nor compulsory. IMBE might engineer people who will be more moral than they otherwise would have been. It is directed to moral enhancement of our offspring. It will be shown why a combination of VMBE and IMBE might be the best option humans have to become better.

References

Doyle, M.W. 1983. Kant, liberal legacies, and foreign affairs. *Philosophy and Public Affairs* 12(3):205–235
Harris, J. 2011. Moral enhancement and freedom. *Bioethics* 25 (2): 102–111
Harris, J. 2016. *How to be good*. Oxford: Oxford University Press

Kant, I. 1907. *Religion innerahalb der Grenzen der bloßen Vernunft* (*Religion Within the Boundaries of Mere Reason*). Ausgabe der Preußischen Akdemie der Wissenschaften (Ak. 6:3–202)
Lewis, C.S. 1943. *The abolition of man*. Oxford: Oxford University Press
Persson, I., and J. Savulescu. 2008. The perils of cognitive enhancement and the urgent imperative to enhance the moral character of humanity. *Journal of Applied Philosophy* 25/3:162–177
Persson, I., and J. Savulescu. 2011. *Unfit for the future*. Oxford: Oxford University Press
Persson, I., and J. Savulescu. 2014. Should moral bioenhancement be compulsory? Reply to Vojin Rakic. *Journal of Medical Ethics* 40 (4): 251–252
Persson, I., and J. Savulescu. 2017. The duty to be morally enhanced. *Topoi*. Online first:12 April 2017. Website: https://link.springer.com/article/10.1007/s11245-017-9475-7
Pinker, S. 2011.*The better angels of our nature*. New York: Viking Books
Rakić, V. 2014. Voluntary moral enhancement and the survival-at-any-cost bias. *Journal of Medical Ethics* 40(4):246–250
Rakić, V. 2017. Moral bioenhancement and free will: continuing the debate. *Cambridge Quarterly of Health Care Ethics* 26(3):384–93
Siverman, Ed. 2004. The 5 most pressing ethical issues in biotech medicine. *Biotechnology Healthcare* 1(6):41–45

Contents

1 **Enhancing Performance** 1
   1.1 Bio-conservatives, Bio-liberals and the Argument of Nature 1
   1.2 Arguments Other Than Nature 6
   1.3 Is Moral Bioenhancement a Solution? 9
   References 10

2 **Morality and Moral Bioenhancement** 11
   2.1 Love and Morality 11
   2.2 Comprehending Goodness and Being Good 15
   2.3 Moral (Bio-)enhancement 17
   Reference 18

3 **Support and Opposition** 19
   3.1 Moral Bioenhancement—Categorical Support 19
   3.2 Moral Bioenhancement—Hypothetical Support 24
   3.3 Moral Bioenhancement—Hypothetical Opposition 28
   3.4 Moral Bioenhancement—Categorical Opposition 33
   References 34

4 **Categorical Opposition to MBE: Harris Wiseman** 37
   4.1 Where CMBEO and VMBO Disagree 38
   4.2 The Groundwork of the Disagreements 41
   References 47

5 **Realistic Means of Enhancing Morality and Why Compulsory MBE is Ineffecive** 49
   5.1 Substances/Medication and Technologies That can Morally Bioenhance Human 50
   5.2 Compulsory MBE is Ineffective 56
   References 59

6 **Voluntary Moral Bioenhancement and Happiness as Its Grounding Rationale: The Best Option on Offer** 61
   6.1 Freedom and Survival 61

6.2 Happiness: The Reason for Being Good? 66
　　6.2.1 Evidence 69
6.3 Goodness, Happiness, Self-interest and MBE Incentivization
　　in Synergy 71
References 74

**ADDENDUM: Combining VMBE and IMBE—A Future Beyond
the Garden of Eden** 75

Chapter 1
Enhancing Performance

> *Whatever you are, be a good one.*
> —Abraham Lincoln

1.1 Bio-conservatives, Bio-liberals and the Argument of Nature

Humans have always aspired to transcend what nature has endowed them with. A lot of scientific achievements have been motivated by it, while warnings about its dangers have inspired many philosophers, scientists and artists to turn to this theme. Icarus is reported to have tried to fly, but paid with his life for this blasphemic desire, that is, the longing to become more powerful than God planned man to be. Goethe's Faustus went even so far as to conspire with Satan in order to obtain the wisdom God did not envision for humans. This resulted in Faustus' insanity. Both Icarus and Faustus paid a high price for their ambition: death in the case of Icarus and severe disappointment in what Satan was able to offer, madness and self-annulment in the case of Faustus.

As the motif of performance enhancement is an important part of humanity's cultural heritage, before discussing the ethical aspects of moral enhancement in detail, the morally relevant features of human enhancement in general will be addressed. Performance enhancement will receive special attention.

+++

Those who insist on the dangers of humans acquiring the powers nature has not endowed them with are labelled as "bio-conservatives". Here I will enter a personal note in this book. I never had the opportunity to meet more than one or two really influential bio-conservatives at one place. In November 2011 a number of my colleagues helped me in organizing in Belgrade (Serbia) the conference "New Perspectives in Bioethics". It was a rare occasion on which the scientific community in Belgrade had the opportunity to meet at one place people like John Harris, Ingmar Persson, Tom Douglas, James Hughes, Katrien Devolder and other scholars whose focus is largely on the issue of human enhancement.[1] One of the controversial issues that arose almost immediately after the conference has been opened was the question whether it is easier for someone to endow in a short time span many humans with a major good or to inflict major harm on a large number of people in a similarly short period of time. John Harris and Ingmar Persson started an intense debate on this, while Tom Douglas joined the discussion a few moments later. The deliberation went in the direction of the best ways how to protect humanity against ultimate harm. What struck me during the whole debate was that nobody defended anything even remotely resembling the bio-conservative position. I was of course aware that most of my invitees at the conference were bio-liberals who supported human enhancement, but I wasn't aware that nobody in the conference room would bring into question the usefulness and moral correctness of enhancement.

In the evening hours some of the conference participants went to have dinner at the Belgrade restaurant *Kolarac*. Ingmar Persson took a seat in front of me and we started a conversation about his and Julian Savulescu's approach to moral enhancement. I voiced two objections to their approach. First, I objected to their conception of halting technological advances until humans become sufficiently moral to make responsible use of them. In particular, I had a problem with halting cognitive bio-enhancement. One of my arguments against it was that cognitive enhancement might contribute to moral enhancement and consequently to lowering the likelihood of humanity destroying itself. Namely, the more humans know, the less prejudices they have, the more moral are they likely to be. Hence, they will be less prone to inflict ultimate harm on humanity—according to the criteria of Persson and Savulescu themselves. Second, I objected to making moral enhancement by biomedical means compulsory. My stance was that by making it compulsory humans would be deprived of an essential foundation of their humanness: their freedom.

I remember that Ingmar argued that my first objection was based on a misunderstanding. Ingmar insisted that he and Julian Savulescu did not think that technological advances, including cognitive enhancement by biomedical means, ought to be halted until humans become sufficiently morally enhanced. We discussed this issue at some length. Unfortunately, we had somewhat less time to discuss my second objection, that is, the issue of whether compulsory MBE would diminish our freedom of the will. Ingmar did however spend some time defending compulsory MBE during our dinner. It was my objections against this stance that have turned out to be the crux of what I have written in the time to come about the position advocated by Persson and Savulescu. Some key ideas on this issue that I developed later were initiated by my conversation with Ingmar at *Kolarac*.[2]

+++

---

[1] A few months after that conference, the Belgrade based Center for the Study of Bioethics has been founded.

[2] After our dinner I took a walk with John Harris through Vuka Karadžića Street. This street, located in the very city center of Belgrade, has become a few years later the seat of the Center for the Study

1.1 Bio-conservatives, Bio-liberals and the Argument of Nature

In what follows I will argue, in opposition to the bio-conservative position, that performance enhancement is not morally controversial *in principle*.

The main objections against performance enhancement are the following: it is unnatural; its use is a form of cheating; its use is not necessarily cheating, but it still places its users at an unfair advantage vis-a-vis those who don't use it; it might be unsafe.

In this chapter I will argue against all these objections. My main, though not exclusive focus will be on cognitive bioenhancement, notably pharmaceutical cognitive enhancement (PCE). The reason for added attention to PCE lies in the fact that currently available non-traditional cognitive enhancements are largely pharmaceutical in nature. It will follow from my argument that performance enhancement, specifically PCE, is not morally controversial *in principle* and that it is morally plausible to use and to prescribe, under certain circumstances, effective and safe cognitive enhancers, including PCEs. As the first objection (performance enhancement is wrong because it is unnatural) is the most exploited one, the bulk of my attention will be directed to that objection.

Bio-conservatives traditionally maintain that performance enhancement is morally impermissible *in principle*, because humans are not supposed to alter what God has ordained or nature has shaped. (P)CEs are therefore also morally impermissible *in principle*. Conversely, by arguing that the use of (P)CE is not morally controversial *in principle*, the traditional bio-conservative claim is being directly opposed.

Bio-conservatives argue that man is not supposed to "play God". "Playing God" implies here that those who are guilty of this charge are assuming divine powers without possessing divine wisdom. It is alleged to be the "hubris of acting with insufficient wisdom" (Kass 2003: 287). If "God" is replaced by "nature", the bio-conservative charge becomes that it is morally wrong to interfere with the naturally given. This begs the question what kind of morally dubious powers we are assuming in that case. I will argue that there is nothing morally dubious *in principle* in interfering with the naturally given. Before coming to that, a few words will follow about the history and present state of the bio-conservative argumentation.

In general, bio-conservative arguments against performance enhancement used to be more ferocious than they currently are. Similar to many other new technologies, performance enhancement by biomedical means has been strongly criticized as soon as it became available. Let us recall some of the arguments we used to hear during the previous decades.

Kass (2002), Fukuyama (2003) and Sandel (2004), but also Habermas (2003), focused their misgivings about enhancement in relation to the perils it allegedly poses to human nature, dignity and freedom. Kass (2002) argued that contemporary bioethics overlooks the danger of human dignity being downgraded by enhancement. Fukuyama (2003) also alleged that human dignity can suffer adverse consequences if human nature is altered by enhancement technologies. Habermas (2003) warned

---

of Bioethics. At that time I didn't even plan to found the Center. The idea that a few years later it would be seated in Vuka Karadžića Street wasn't even in my remotest imagination.

about the impact of enhancement technologies on our conceptions of what it means to be human and to be moral. According to Habermas, genetic alterations bring into question our autonomy and standing as moral agents. Sandel (2004) critically assessed our need to control nature in the case of reproductive technologies—a need Sandel believed can lead to our inability to appreciate life as a gift (see Rakić 2012).

Annas (2000) and Elliott (2003) critically assessed the utilization of enhancement technologies in general. Annas argued that the misuse of science is neither limited to the Nazi past, nor to nuclear, chemical and biological weapons. Genetic engineering, namely, threatens the survival of the human species as well. Elliott addressed the alleged American fascination with enhancement technologies (Botox, Viagra, mood lifters). He believed that, although they are being used as remedies against social phobias, they ironically appear to adversely affect their users' self-consciousness (see Rakić 2012).

All in all, the bio-conservative claims against performance enhancement come down, in one form or another, to the position that humans should avoid interfering with what nature has endowed them with. But there is no proof of a causal link (arguably not even of a positive correlation) between enhancement (in general) and diminished dignity, freedom, autonomy or self-consciousness of the users of enhancement technologies. Similarly, there is no evidence that there is an inherent quality of reproductive technologies that interferes with the "ability to appreciate life as a gift". Hence, the two best options in bio-conservative argumentation apparently remain to either show that there is something wrong with enhancement *in principle* or to search for innovative arguments against it.

The use of PCEs and other performance enhancers is indeed "unnatural" in the sense that it can extend performance beyond what nature has endowed humans with, and consequently transcend species-specific functioning. The question is however what is morally controversial about this. Is it merely the fact that performance enhancement is "unnatural" (argument A) or is it this fact in combination with the fact that we deal with enhancements (argument B)? None of the two arguments holds for the following reasons.

Argument A. Various natural phenomena inflict harm on humans who wish to avoid, bypass or eliminate them. Examples include earthquakes, tsunamis, avalanches, covid-19 and other severe diseases. Natural phenomena can be a source of enjoyment for humans, as well as a source of suffering. John Harris has noted various times that nature is morally indifferent. If we conceive morality in a utilitarian fashion and define it as the maximization of happiness, this is undoubtedly true. But even if we don't understand morality in that way, it is still beyond dispute that nature can be both a source of happiness for humans, as well as a source of their suffering. Hence, there is nothing wrong *in principle* in maximizing our happiness by utilizing "unnatural" means to achieve something. If it would be wrong *in principle*, we should ban drugs (unless they are produced by the natural environment) and vaccines, but also cars, trains, refrigerators, vacuum cleaners, phones, TVs, computers, electricity in general. As has been pointed out by various scholars who have dealt with this issue, the lives of almost all human beings are deeply unnatural and bear little resemblance to our species' "natural" state. It is indeed immensely difficult to use the argument of "naturalness" in order to show that a certain intervention is morally wrong.

Argument B. It is however possible to argue that the objection based on the argument of "naturalness" should apply only to enhancement technologies. Hence, only "unnatural" *enhancements* are morally dubious. The treatment of diseases by "unnatural" means would on the other hand be morally uncontroversial. However, the possibilities that genetic technologies appear to offer in preventing rare diseases, cancer, HIV/Aids (e.g., CRISPR-Cas 9 as the one with most publicity in recent times), the engineering into cells of resistance to cancer and HIV/AIDS, as well as the well-known traditional vaccines, are obviously enhancements, as they make humans go beyond "species typical functioning". They are therefore also "unnatural". But not many reasonable people will *prima facie* argue against their moral appropriateness, as they are designed to save lives or contain dangerous/debilitating diseases. The question is what the differences are, in moral terms, between these enhancements on the one hand and on the other hand novel PCEs that can improve memory and concentration. Both are enhancements and both are unnatural. Even if saving lives might be a motive that is morally superior to cognitive enhancement, preventing diseases that lower our quality of life is not something that is by definition morally superior to cognitive enhancement. It is therefore warranted to conclude that the "unnaturalness" of cognitive enhancement technologies does not make them morally inappropriate *in principle* vis-a-vis other "unnatural" enhancement technologies that might preserve our health/lives or improve our quality of life.[3]

All in all, the argument that the use of performance enhancement by "unnatural means" is morally impermissible (or even dubious), because it is "unnatural", is not a sufficiently strong one. In one way or another, this position is conveyed by bio-liberals.

Unlike bio-conservatives, bio-liberals maintain that an upgraded human is a better human. Consequently, enhancement is morally permissible, possibly even our moral duty. Moreover, refusing enhancement to those who wish to subject themselves to it decreases their freedom. Conversely, the enlargement of human possibilities amplifies the number of options we have in our lives, enabling us, among other advantages we acquire, to learn, produce and earn more. Hence, it increases our freedom. It is therefore not bio-technologies but the state that is the primary potential culprit for infringing upon our freedom and for denying our pursuit of happiness through self-improvement (Rakić 2012).

Since human well-being is essential, it is not only the treatment and prevention of disease that is relevant, but also the enhancement of human possibilities. Hence, if it is our duty to treat and prevent disease, it is, according to some bio-liberals, also our duty to intervene in what is given to us by nature in order to provide an individual with the best prospects for having the best possible life (Savulescu 2006: 525).

---

[3] Some supporters of enhancement technologies even argue that it is not only morally permissible to use enhancement technologies to make people more healthy, longer-lived and smarter, but that we are morally obliged to do so (e.g.: Harris 2010 or Savulescu 2007).

Arguments in favor of the moral *permissibility* of enhancement can be found in Agar (2003). Agar argues that enhancement should be permissible but not mandatory. In that context he distinguishes between authoritarian and liberal eugenics. The former advocates a monistic outlook on human excellence, while the latter embraces a pluralistic perspective. Such a perspective is marked by an absence of a conception of a single desirable genome and consequently an absence of compulsion in order to arrive at such a conception. In that sense liberal eugenics ought to be disassociated from the eugenic practices of the Nazis (Rakić 2012).

Agar's arguments in favor of the moral *permissibility* of enhancement are radicalized by Harris and Savulescu in that they advanced the thesis that we have a moral *duty* to enhance. According to Harris (2007) it is not only morally acceptable to use genetic technology to make people healthier, longer-lived and more intelligent, but in most cases it is also our moral duty. A radical augmentation of our mental and physical capabilities will influence the very course of evolution: new types of regenerative medicine appear to open up the possibility of human tissue to repair itself, techniques are becoming available that can radically extend life expectancy, while new drugs can improve concentration and memory and enable us to function successfully with less sleep. Harris does not see, rightly so, a morally relevant difference between enhancements that will make us healthier and longer-lived and enhancements that will upgrade our cognitive capacities. Hence, we should enhance ourselves in almost any way we desire.

Savulescu (2002) argues that parents have the moral right to decide about their children's genes that is similar to the right they have regarding their rearing and education. There are two reasons why procreative liberty is to be extended to enhancement. First, since the raising of children is a private matter and parents must bear most burdens of having children, they have a justifiable interest in the nature of the child they are bringing up. Second, as it is only through "experiments in living" that people find out what they think is best for them, diversity in choice is essential to discovering which lives are optimal for human beings (Ibid., 526, 527).

Everything considered, what remains is the difference between a naturally given and an improved human being. If we have to make a selection between these two, argue bio-liberals, it is our duty to select the latter. We are obliged to try to become better, an enhanced human being is a better human being, and consequently we have a duty to enhance. Moreover, society has a duty to provide us with the best opportunities for attaining enhancement.

1.2 Arguments Other Than Nature[4]

What follows is a discussion of a few additional arguments against performance enhancement. Occasionally it is argued that performance enhancement, specifically the use of PCEs, is a form of cheating. That argument is wrong. Cheating is relative

---

[4]For the paragraphs that follow, consult also Rakić (2017).

to a certain rule. As there are no widely accepted rules in place when the use of PCEs is concerned, there is nothing to cheat on. Only if a certain institution does have such rules can we talk about cheating. For instance, if there is a rule prohibiting those who take certain exams to use PCEs, the utilization of PCEs would mean breaking rules. In that sense it would be a form of cheating. Those who break rules would then rightfully get kicked out of the exam. That is precisely what should happen to those who break rules.

This is also an essential difference between the use of PCEs and the utilization of doping in sports. In sports it is clearly defined what the rules are and thus what cheating means. These rules might be a cultural construct, but they are rules that are to be respected in order not to be accused of cheating.

Another argument against the use of PCEs is that its utilization allegedly puts its users at an unfair advantage vis-a-vis those who don't use them. In order to answer this argument we ought to consider several types of fairness. First, is it fair that a natural advantage in the cognitive realm is likely to place one in a better position vis-a-vis those who don't have such an advantage, while the artificial creation of such an advantage is unfair? In sports this is indeed the case: if you have a natural advantage, you are deemed to have a fair advantage; if you have created this advantage artificially, in particular by using performance enhancing substances, you are considered to having acquired an unfair advantage. But this rule is also a cultural construct that can be brought into question. One of the reasons to bring it into question is that it is based on the distinction between "natural" and "unnatural" - a distinction the arbitrariness of which has already been discussed. Moreover, in the case of cognitive enhancement there is even no widely accepted informal agreement of that type—an agreement according to which the exploitation of "natural" inequalities is morally justified, while the exploitation of artificially created inequalities is morally impermissible.

Second, is it morally justified that those who are socially and financially better off have an advantage in getting access to PCEs? Most likely it is not. It can be argued that acquiring the financial means of obtaining access to PCEs might be the result of hard work and self-sacrifice, but we all know that the acquisition of wealth is not always the result of that. Moreover, huge discrepancies in wealth are frequently considered to be morally dubious. This is however not a phenomenon that affects only the use of PCEs. Unequal distribution of wealth makes it possible to those who are better off to acquire all kinds of advantages for themselves and their offspring. There is no reason to single out PCEs, ban their use, and say that they are banned because they put those who can afford them at an advantage. What about all other advantages the better off have in their lives? If we wish to address the issue of unequal distribution of wealth we should do that at a different, more general level than the level of access to PCEs.

Savulescu (2007) is also not favorably disposed to the bio-conservative argument pertaining to equality. The fear of a two-tier society of the enhanced and the unenhanced is not justified, argues Savulescu, because allowing enhancement will help the ungifted approach the gifted. The lottery of nature might therefore be equally or even less fair than enhancement. Furthermore, how well the lives of those who are deprived go depends not on whether enhancement is allowed, but on the social institutions we have to protect the underprivileged and provide everyone with a fair chance in life (Ibid., 530). Savulescu argues that those who oppose enhancement are guilty of a "crude form of social determinism": they envision undesirable social outcomes if enhancement is allowed, although it is within our power to try to avoid such outcomes—precisely by enhancement that has the potential to reduce inequality (Savulescu 2006: 336).

The arguments regarding enhancement and inegalitarianism can even be radicalized. There is reason to assert that because humans are born unequal, nature is inegalitarian. Hence, if we consider egalitarianism as an important moral value, we have to conclude that nature is immoral. The thesis of John Harris that nature is not moral, but morally indifferent, would in that case have to be reshaped into the view that nature is immoral. As a matter of fact, it might be argued that either egalitarianism is not moral or that nature is immoral. If we replace nature by God, we would have to conclude that either egalitarianism is not moral or that God is immoral!

Furthermore, it is possible to minimize real or perceived injustices of the type we discuss here by developing health insurance plans that cover the use of PCEs. Such plans frequently do cover the use of Viagra. In the majority of cases Viagra is being used as a performance enhancer (Mehlman 2011: 127). There is no reason to exclude the possibility to have similar policies in the case of PCEs.

Another indictment of PCEs comes down to the point that employers and other authorities may be misled by the initial performance of those who use them. For example, if a candidate for a job enhances her performance during the job interview, while after the interview her functioning returns to her functioning without the help of PCEs, the employers might expect more from her than she actually would be capable of on the basis of her unenhanced abilities. The employers would be misled. Consequently, they could make a wrong choice at the detriment of various competitors or/and at their own detriment.

In certain competitive contexts such as the one that has just been stipulated, the cognitively enhanced individual would at one point face the dilemma whether to admit to her employers what the limits are of her unenhanced performance *or* to continue to use PCEs. The latter option might in certain cases be detrimental to her health.

On the other hand, in all competitive contexts the issue of "misled authorities" can be addressed by the disclosure requirement: the obligation of competitors to disclose information about the use of PCEs. The disclosure requirement might infringe to a certain extent on our privacy, that is, intrude into our seclusion, but the consequence of a lack of disclosure can lead to unfair outcomes in competitive contexts. The values of privacy and fairness ought to be balanced out in this context. The disclosure requirement might also not be easily enforceable (competitors might lie about it). Nonetheless, if it were discovered that a competitor had been untruthful, this would amount to rule breaking, cheating, and thus be sufficient reason for the competitor to be banned from the competition, to be fired or to suffer other drastic consequences.

The issue of safety is another argument against the use of PCEs. But what is "safe"? Some PCEs are undoubtedly unsafe for some individuals. Not only do people react differently to different substances, but some individuals might be tempted to exceed the prescribed dose of a PCE. Knowing that a certain PCE gives them an advantage vis-a-vis other competitors, possibly suspecting that other competitors take PCEs, some might decide to increase the doses their physicians have prescribed. Depending on how far they are willing to go in increasing the dose, as well as on how their organisms react to this increase, they might bring their health into jeopardy.

Physicians who prescribe PCEs have to take into account various issues that are medically relevant for their patients. They range from the effectiveness and safety of the substances they prescribe to how each patient's organism reacts to them. They have to be well informed about the effects and side-effects of various PCEs, of the medical history of their patients and whether some PCEs are counter-indicated in certain cases. If physicians properly take into account all elements that are relevant for coming to a decision in the case of prescribing a PCE, they have addressed the issue of safety in an appropriate manner.

1.3 Is Moral Bioenhancement a Solution?

In conclusion, the main objections against the use of performance enhancement, specifically PCEs, are not sufficient for banning PCEs, for interested individuals *prima facie* deciding not to use them or for physicians *prima facie* deciding not to prescribe them. The discussed reservations do not warrant banning PCEs *in principle* or any other type of performance enhancement *in principle*.

Returning to the opening paragraph of this chapter, if performance enhancement is indeed not morally controversial *in principle*, why have there been so many warnings in our cultural heritage about the dangers of human enhancement? Have they been merely conservative prejudices or is there more to them? I think that there is more to them. Serious warnings dealt with the issue of whether humans are capable of judging their capacities and whether they are morally apt to enhance these capacities. We have mentioned the tragic fates of characters such as Icarus and Faustus. But their cases do not discard performance enhancement in principle. They merely address wrong ways of enhancing humans.

Taking into account all reasons discussed in this Chapter that demonstrate that cognitive and other performance enhancements are not morally controversial *in principle*, and even if we conclude with Savulescu that enhancement "expresses the human spirit" and that "to be human is to be better" (Savulescu 2006: 531), the question still remains whether we have the moral capacity to decide in which ways to enhance ourselves. And if we do not have such a capacity, is *moral* bioenhancement a possible solution?

References

Agar, Nicholas. 2003. *Liberal Eugenics: In Defence of Human Enhancement*. Oxford: Blackwell.
Annas, George. 2000. The Man on the Moon, Immortality and Other Millennial Myths: The Prospects and Perils of Human Genetic Engineering. *Emory Law Journal* 49 (3): 753–782.
Elliot, Carl. 2003. *Better Than Well: American Medicine Meets the American Dream*. New York: W. W. Norton.
Fukuyama, Francis. 2003. *Our Posthuman Future: Consequences of the Biotechnology Revolution*. London: Profile.
Habermas, J. 2003. *The Future of Human Nature*. Cambridge: Polity Press.
Harris, John. 2007. *Enhancing Evolution: The Ethical Case for Making Better People*. Princeton: Princeton University Press.
Harris, J. 2010. Moral Enhancement and Freedom. *Bioethics*. https://doi.org/10.1111/j.1467-8519.2010.01854.x.
Kass, Leon R. 2002. *Life, Liberty and the Defense of Dignity: The Challenge for Bioethics*. San Francisco: Encounter Books.
Kass, Leon R. (ed.). 2003. *Beyond Therapy: Biotechnology and the Pursuit of Happiness*. Washington, DC: The President's Council on Bioethics.
Mehlman Maxwell, J. 2011. Modern Eugenics and the Law. *Faculty Publications, School of Law, Case Western Reserve University*: 1579.
Rakić, V. 2012. From Cognitive to Moral Enhancement: A Possible Reconciliation of Religious Outlooks and the Biotechnological Creation of a Better Human. *Journal for the Study of Religions and Ideologies* 11 (31): 113–128.
Rakić, V. 2017. Cognitive Enhancement: Are the Claims of Critics 'Good Enough'? *Cambridge Quarterly of Healthcare Ethics* 26 (4): 693–698.
Sandel, M. 2004. The Case Against Perfection. *Atlantic Monthly* 3: 51–62.
Savulescu, J. 2002. Abortion, Embryo Destruction and the Future of Value Argument. *Journal of Medical Ethics* 28 (3): 133–135.
Savulescu, J. 2006. Justice, Fairness and Enhancement. *Annals of the New York Academy of Sciences* 1093: 321–338.
Savulescu, Julian. 2007. Genetic Interventions and the Ethics of Enhancement of Human. In *The Oxford Handbook of Bioethics*, ed. Bonnie Steinbock, 516–535. Oxford: Oxford University Press.

Chapter 2
Morality and Moral Bioenhancement

> *I think the devil doesn't exist, but man has created him, he has created him in his own image and likeness.*
> —Fyodor Dostoevsky, The Brothers Karamazov
> *If a man thinks about his physical or moral condition, he generally finds that he is ill.*
> —J. W. Goethe

Why aren't we better? Do we have a problem with understanding morality or don't we want to be better? Before turning to the theme of *enhancing* morality, it is in order to point to a number of issues that are essential for a proper *comprehension* of morality. In this chapter, in the section that follows, the question will be raised what it is that makes humans less good than they *can* be and what they may do in order to become better than they are. It will be argued that love is a key component of goodness, as well as an act of our will. In the section after that, "goodness" will be specified and the concept of the "comprehension-motivation gap" will be introduced in some detail. Finally, the issues of freedom, morality and MBE will be addressed.

2.1 Love and Morality

> *Hell is yourself and the only redemption is when a person puts himself aside to feel deeply for another person.*
> Tennessee Williams
> *What is hell? I maintain that it is the suffering of being unable to love.*
> Fyodor Dostoevsky, The Brothers Karamazov
> *Love loves to love love*
> James Joyce, Ulysses

The virtue of altruism occupies a key place in modern approaches to morality, apparently irrespective of culture. Altruism is closely related to empathy. It is a common habit of our modern minds to consider it as essential to morality, in addition to the manifestations of the virtue of justice. It can be argued that empathy as the capacity

© The Author(s), under exclusive license to Springer Nature Switzerland AG 2021
V. Rakić, *How to Enhance Morality*, SpringerBriefs in Ethics,
https://doi.org/10.1007/978-3-030-72708-6_2

to understand and feel from the point of view of other people what they are experiencing is at the basis of the "Golden Rule"—to wish the same to others as one wishes to herself. As the "Golden Rule" is an essential moral value, largely irrespective of culture, empathy and altruism are central to morality in general.

The literary heritage of humanity is full of descriptions of the moral heroisms and moral downfalls of the characters it has been portraying. Few authors have depicted the moral downfalls of humans in such a blatant form as William Shakespeare has. Macbeth is obsessed with power and ambition, Shylock and the feuding families Capulet and Montague with revenge and power/domination, Richard III with envy resulting from his physical deformity, Othello with jealousy. Shylock and Othello were additionally being subjected to religious and racial bigotry.

A lot of passions of these characters appear to be grounded in hatred. Their follies result in grave misdeeds, which are however sometimes brought into question by repentance, mercy, as well as by compassion and empathy. So says Macbeth to Macduff: "Of all men else I have avoided thee: But get thee back, my soul is too much charged with blood of thine already". These words have no impact on Macduff whose agony caused by Macbeth, his hatred towards him, desire for vengeance and for power dominate everything else. So he replies: "My voice is in my sword: thou bloodier villain than terms can give thee out".

The pain that the Capulets and Montagues have inflicted upon each other does however result in some good. They abort their enmity after the deaths of their adored children Romeo and Juliet, whose love for one another could not be sustained because of family feuding and hatred. This makes Escalus, the Prince of Venice, utter these words to the Capulets and Montagues in the final part of the play: "See, what a scourge is laid upon your hate, that Heaven finds means to kill your joys with love."

The theme of hatred being defeated by love is a common one in humanity's cultural heritage. This book began with the *1001 Nights of Sheherezade*. Love defeats Shahryar's pathological hatred toward women. Sheherezade is the instrument of love and Shahryar's moral enhancement. The love of Romeo and Juliet and their death are the instruments of "Heaven" to "kill hatred's joys with love".

Love is a mindset that involves caring, deeply and personally, about the objects of love for their own sake. Love is largely the groundwork of morality. The moral principle to desire things to happen to others that one desires to happen to herself is an essential feature of the concept of love. The wish that justice be done to those who made innocent people suffer is partially based in love for those people, and perhaps for humans in general. As love is an important human inclination, and as morality is intertwined with love, it is even more difficult, for almost any human, not to care about morality.

Love often appears to be the rationale of morally laudable actions, while hatred frequently manifests itself through immoral behavior. Altruism as the basis of the "Golden Rule" is also largely rooted in love. Moreover, love appears to be less convoluted in its manifestations than hatred. We mentioned in the foregoing paragraphs some of the many manifestations hatred can take. Love, on the other hand, is more forthright. To a significant extent, it manifests itself in simply *being good*

2.1 Love and Morality

(to others and to oneself). Kindness, helpfulness, generosity are some of its manifestations. Hatred displays itself in multifarious, intricate ways, taking the form of envy, physical or psychological violence affecting those who are weaker than we are, ruthless ambition, bigotry, murder, (blood) revenge, obsession with power leading to humiliation of people, molestation...

The relative frankness of love and convolutedness of hatred might have been one of the reasons for Tolstoy beginning his *Ana Karenina* with the famous illustrious words "All happy families are alike; each unhappy family is unhappy in its own way." Assuming that Tolstoy's observation is correct, why is that so? The reason might very well be contained in happy families living in frank loving relationships, while the unhappy ones suffer from very different kinds of vicious interactions: selfishness, jealousy, rivalry, envy, improper relations with children, dislike of in-laws, dislike and possibly hatred of the spouse because of financial issues, betrayed emotions or competition, as well as other types of morally controversial feelings, dispositions and actions. If love marks interactions in happy families, these interactions are morally superior to the ones of the unhappy families. The relationships in unhappy families are not always marked by hatred or hatred only, but by various other morally controversial interactions. The fact that love is rather forthright, while hatred is complex in its manifestations, might be the reason why unhappy families are unhappy in specific ways. Love apparently contributes to making us both happy and moral in a very special, but honest way. Hatred seems to have a plethora of bizarre manifestations, but with one overarching feature: in most cases it leads to unhappiness and moral decline. Tolstoy's unhappy families seem to be an illustration of that.

It appears that an effective way to morally enhance humans is to enhance their capacity to love. But that is not an easy task. Although moral education might help, it proves not to be sufficient. MBE is also powerless in that sense. Substances such as oxytocin may augment our empathy for some time, but empathy is not love. It might be argued that it is a manifestation of love, but not love itself. Moreover, love is an act of our volition; empathy is not. MBE has only the potential to help in enhancing the manifestations of love. It cannot create or augment love itself. But do these manifestations of love have the potential to make us love or love more than we currently do?

It might be argued, cogently, that it is not worth bothering whether you truly "love" your neighbour, but that you should act as if you did. In due course, you will find out that you have come to love him more. Your manifestation of love has become true love. Similarly, if you make a habit of ill-treating someone you dislike, you may find out disliking him more than you did before.

We can speculate about the reasons behind this psychological mechanism. A sound explanation could be that people tend to search for justifications of their behavior. That does not only mean justifying it in front of others but also in front of their own eyes. In order to feel good about ourselves, we ought to legitimize our behavior. Hence, if we are good to someone, we can easily justify our behavior by our love for that person. If we are bad to someone, our badness can be legitimized by our dislike for that person. Moreover, we can rationalize our love or dislike by starting to believe that someone really is good or bad.

Our manifestation of love/dislike for someone might very well lead to loving/disliking that person or to loving/disliking her more than we did before. For example, certain racial or ethnic minorities have been mistreated because they were hated. But afterwards they could have been hated even more because they were mistreated. The more vicious you are, the more you will hate; the more you hate, the more vicious you become. In that way, mistreatment and dislike operate in a circularly supportive fashion. Conversely, kindness (acting as if you love) and genuine love operate in the same way.

In colloquial jargon love is frequently confused with "being in love", or with the sentiments we have for our close relatives (children, parents, siblings etc.)/friends, or with "liking" someone. But I use here a different conception of love. According to this conception, we are not kind and loving to those who surround us because we are "in love" with them, or because they are our close relatives, or because we "like" them. We are kind and loving because we *decide* so. We decide to accept them as we accept ourselves. Accepting them does not imply "liking" them. It implies wishing them the best. It is similar to what we wish ourselves: the best—in spite of the fact that we may not "like" certain features of our characters. Comparably, we may not "like" certain people, but "loving" them means accepting them as they are and wishing them the best.

This notion disassociates love and affection. Feeling affection for someone is a sentiment. Love, on the other hand, is not a sentiment but an act of the will. Moreover, it is related to morality. Some people are more affectionate than other people, but that does not make them more moral. What does make someone more moral is her decision to act as if she loved those surrounding her. Defined in this way, love is based on our voluntary decision to love. It resembles the morality of our behavior: we voluntarily decide to behave in a more or less moral way.[1]

Both morality and love are acts of our will. Love for one's children, spouse, parents, siblings, friends or nation are types of love that can be called *sentimental*. They are based in our emotions. Love for all people is however based on a decision to love all humans. In that sense it is not sentimental, but rather *volitional love*.

Morality operates in a similar fashion. Morality that extends to "an inner circle" of, for example, children, spouse, parents, siblings, friends or even one's nation, can be called *sentimental morality*. It excludes those outside the circle from the moral criteria that apply to those within the inner circle. Morality extending to all is inclusive and is based not on our sentiment but on our decision. It can be called *volitional morality*.

This links love and morality as volitional acts also from another angle: by acting *as if* we love we will not only become truly more loving, but as a consequence of that, also more moral. If love is at the root of some of the essentials of moral *behavior* (altruism, empathy), manifestations of love might enhance us morally. This is a different wording of what has been argued already: that acting in a kind and loving manner, we tend to truly become more loving. Consequently, if we enhance

---

[1] That is also one of the reasons why only *voluntary* moral enhancement can be genuine moral enhancement. This issue will be discussed later in the book.

our manifestations of love by MBE, this might make us more loving and henceforth more moral. MBE does therefore have the capacity to *indirectly* improve our morality. By improving it, i.e. by making us *act* more morally, we tend to become more moral.

In the following section it will be discussed what "becoming more moral" implies, what it is that we consider as "moral behavior", as "goodness". What does goodness imply? What does it mean "to be good"?

2.2 Comprehending Goodness and Being Good

God has given you one face, and you make yourself another.
William Shakespeare

Man is the 'ethical animal'—ethical in potentiality even if, unfortunately, not in actuality.
Rollo May

The difference between what we do and what we are capable of doing would suffice to solve most of the world's problems.
Mahatma Gandhi

In our everyday conversations it is common that we say things such as "He didn't keep his promise", "He is always there to help his friends/neighbors etc.", "Although her income is modest, she donates a lot of money to the poor in the developing world", "You lied!", "He is cruel to those he considers to be weaker", "This seat is mine, I was there first". In all such statements we apparently refer to a standard that we believe ought to be kept (or think that those who keep it are good people), and about which we believe others also think ought to be respected (or also think, as we do, that good people respect it).

The standard we refer to in the above examples is not a culturally determined convention. It is a standard that is valid irrespective of the culture we are embedded in: there are no and most likely there have never been cultures in which it is considered morally laudable to break promises made to friends, to be cruel to friends who suffer, to be treacherous, to be selfish.... Love for your neighbor is a moral standard that is not merely "different" from the morality of the Nazis. It is a principle that all of us know to be morally superior to Auschwitz. Even many Nazis, deep in their hearts and minds, must have known that.

There might not be many moral standards that are common to almost all humans, but those that are accepted irrespective of culture can be considered to be standards of what we understand to be "goodness". Those who do not think of them as common standards of goodness belong to a tiny minority of people who do not have the capacity to distinguish good from evil. It is a minority that could be compared to the minority of people who do not distinguish colours. The ones are colour blind, the others suffer from moral blindness. Here we will however deal with the vast majority of people who do not suffer from such a disability.

Apart from being aware of the existence of a number of common moral standards, we are also aware of the fact that we frequently fail to behave in line with them. We usually find various excuses for not acting according to those standards. We might break a promise because we thought that we will never end up in a situation in which the validity of our words would be tested, we might have been unjust to our children because we were nervous, we might have grabbed the only remaining seat in a bus in front of an elderly woman because we were tired, we might not donate to charities because we forget to do it…Nonetheless, although we are very good in justifying our moral inadequacies, at bottom we know how frequently we are not up to the moral standards we accept on a cognitive level.

It appears therefore that we accept certain common moral standards on a cognitive level, but that at the level of our behavior we frequently fail to follow them. Hence, there is gap between what we do and what we believe is morally right to do. There is a discrepancy between knowing the good and acting good.

- I call this the "comprehension-motivation gap": we comprehend what is morally right, but our motivation is sometimes not strong enough to act in line with this comprehension.

Apart from the objection that these moral standards are *merely* cultural constructs—an objection that does not hold water because some of those standards are universal—another objection can be made as well. It might be asserted, namely, that the common moral standards are values that humans have accepted in order to maximize their chances of survival, either as individuals or as a species. Hence, the objection goes, we deal here with an instinct that we confuse with a moral standard. If we see that a man who is unknown to us is trapped in a burning house, we might feel two different urges or instincts. One is to help him and possibly save his life, the other is to keep ourselves out of danger. The first urge would amount to our care for the survival of our species, the other to self-preservation. Our moral reflection would however tell us that the right thing to do is to help the man who is in trouble, provided that the likelihood is sufficiently high that we don't die in the fire as well. In other words, moral reflection measures two different instincts by a certain moral standard. Such a standard cannot be one of the two instincts. It is something different: a criterion by which we measure our instincts, as well as our behavior in general. Such a criterion cannot be the same as that which is being measured by it. The common moral standards that are being discussed here are therefore neither cultural constructs nor instincts.

In sum, human beings are faced with the following major predicament in their existence as moral beings: (1) they comprehend that they ought to behave according to certain common moral standards; (2) sometimes they prove not to be motivated enough to act in line with this comprehension. In other words, humans know what goodness is, but they are not always good.

Having made a number of key points about the role of love in morality, goodness and the comprehension-motivation gap, the focus will become now the contemporary debate on the enhancement of the manifestations of morality, i.e. that what is mostly being addressed in MBE literature.

2.3 Moral (Bio-)enhancement

Freedom is central to the human conception of morality. If we are not free, we might perform moral acts unwillingly (if we are compelled to perform them) or even unknowingly, but that does not make us moral. Only the crudest form of consequentialism would applaud as morally laudable those nice deeds we have been coerced into carrying out. In order to be genuinely moral we ought to have moral dispositions that guide us to certain actions. We have to be free to choose how much we will keep our actions in line with our sense of right and wrong. We have to decide ourselves how to address the comprehension-motivation gap in our personal lives.

This does not imply that humans necessarily always like to be free. When the Grand Inquisitor addressed Jesus Christ who briefly appeared in Dostoyevski's "Legend of the Grand Inquisitor" in *The Brothers Karamazov*, he was highly upset because Jesus likes humans to remain free to sin, while the Grand Inquisitor made people happy, he believed, by taking away their freedom.

Jesus did not accuse the Grand Inquisitor in the first place for establishing an inegalitarian, corrupt and criminal order. The greatest problem Jesus had with the Grand Inquisitor had to do with his attitude toward freedom. The Grand Inquisitor pretended to be the decision maker in issues pertaining to the redemption of humans. Humans didn't have to care about how to use their freedom to behave morally. The Grand Inquisitor did that instead of them. Humans were not free to fall, as the Grand Inquisitor pretended to do all the hard work for them in order *not* to let them fall.

The Grand Inquisitor asked Jesus why he came to exert his disturbing influence and concluded that it were better if he had never come in the first place. He hatefully accused Jesus of trying to make humans unhappy again by returning their freedom the Grand Inquisitor had taken away for the sake of human happiness. Jesus responded by kissing the Grand Inquisitor on his "bloodless lips", subsequently disappearing from the scene.

This legend has been told by Ivan, one of the Karamazov brothers. Ivan instigated his father's servant and son (Ivan's half-brother) Smerdyakov to kill his lord/father. He did that primarily on the level of intellectual prescriptions, derived from the philosophy of the Grand Inquisitor. The Grand Inquisitor, impersonating God, instructed his subjects that everything is morally allowed—lest not banned by him (symbolizing any Earthly power).

The motives of Ivan and Smerdyakov were, most likely, hatred toward their father, as well as potential financial gain. The murder led to Smerdyakov committing suicide some time after his crime and Ivan developing a mental condition resembling reactive psychosis. According to the criteria of the Grand Inquisitor, Ivan hasn't committed any wrong: no freedom and no God (only the Grand Inquisitor)—everything allowed (lest not banned by the Grand Inquisitor). Ivan's mental illness appears to testify however that Ivan was intimately not fully supportive of the philosophy of the Grand Inquisitor. Serving the opportunistic prescriptions of the Grand Inquisitor to forfeit his freedom (by transferring it to the Grand Inquisitor) led Ivan to despair.

Dostoyevski's tragic character apparently didn't truly believe that forfeiting one's freedom by delegating it to a higher Earthly power is sufficient for being good and happy.

In MBE literature, a concept that resembles Dostoyevski's Grand Inquisitor is the "god machine", as advocated by Savulescu and Persson (2012). The "god machine" is designed to disable humans to realize thoughts that it considers as "grossly immoral". Once such thoughts come up in the mind of an individual and she decides to act in line with them, the "god machine" reacts by "deleting" this intention. The human is not free anymore—his freedom being transferred to the "god machine".

Applying the ideology of the Grand Inquisitor to MBE, at least two questions arise. First, will Savulescu and Persson's "god machine" do a better job than the Grand Inquisitor? Second, is MBE in general doing something similar as the Grand Inquisitor or shall we charge only compulsory MBE with that, sparing voluntary, elective MBE from this censure? Or do none of the two MBE options deserve such a harsh criticism? Let us take a closer look now at a number of essential positions on MBE and find out which answers can be found there.

Reference

Savulescu, Julian, and Ingmar Persson. 2012. Moral Enhancement, Freedom, and The God Machine. *The Monist* 95 (3): 399–421.

Chapter 3
Support and Opposition

The positions of some of the most essential MBE scholars will be discussed in more detail now. Their stances regarding MBE will be divided into four groups on the basis of the criteria support/opposition/categorical/hypothetical. First will be discussed those scholars who categorically support MBE: primarily, Ingmar Persson and Julian Savulesu. Second, those who support MBE in the context of certain assumptions: Thomas Douglas and David DeGrazia. Their MBE support is hypothetical. Third, those who hypothetically oppose MBE, that is, in the framework of certain assumptions they make about the usefulness of MBE: John Harris, Nicholas Agar and Robert Sparrow. Fourth, categorical opposition to MBE will be analyzed on the basis of the writings of Harris Wiseman. Much of my focus will be on his arguments against MBE, because he opposes it *in principle*. It is essential to persuasively refute them in order to successfully advocate the type of MBE I advocate, that is, VMBE. At various instances the four approaches to MBE will be compared and contrasted to the concept of VMBE. It almost goes without saying that the selected scholars whose positions will be addressed in this chapter are not the only important ones belonging to each of the four approaches. For the purposes of my argument they are however the most relevant ones.

3.1 Moral Bioenhancement—Categorical Support

In this section the positions of the two most influential categorical MBE supporters will be discussed in more detail: Persson and Savulescu.

Persson and Savulescu

Persson and Savulescu make two arguments that stand out as the most essential ones in the context that is relevant here. First, they suggest something that might be interpreted as advocating the position that cognitive enhancement ought to be halted (or at least slowed down, similar to various other "precarious" scientific advances) until humans have become sufficiently morally enhanced. Second, they promote *compulsory* moral enhancement.

In Persson and Savulescu (2008) it is argued that moral enhancement ought to "accompany" other forms of enhancement, specifically *cognitive* enhancement: "For if an increasing percentage of us acquires the power to destroy a large number of us, it is enough if very few of us are malevolent or vicious enough to use this power for all of us to run an unacceptable increase of the risk of death and disaster. To eliminate this risk, cognitive enhancement would have to be accompanied by a *moral* enhancement which extends to *all* of us, since such moral enhancement could reduce malevolence" (Persson and Savulescu 2008: 166). The argument that cognitive enhancement "would have to be accompanied" by moral enhancement, appears to imply that the latter should be avoided until humans become sufficiently morally enhanced. In the words of Persson and Savulescu: "Therefore, the progress of science is in one respect for the worse by making likelier the misuse of ever more effective weapons of mass destruction, and this badness is increased if scientific progress is speeded up by cognitive enhancement, *until* effective means of moral enhancement are found and applied" (Ibid., 174; emphasis added).

That cognitive enhancement ought to be *preceded* by moral enhancement might also follow from Persson and Savulescu's reference to Lewis's stories and the "Deplorable Word" (a magical curse which will end all life in the world except that of the one who pronounces it).

"If we all knew the Deplorable Word, the world would likely not last long. The Deplorable Word may arrive soon, in the form of nanotechnology or biotechnology. Perhaps the only solution is to engineer ourselves so we can never utter it, or never want to utter it" (Ibid., 175). In other words, we ought to be morally "engineered" so that we will never be able to destroy ourselves by the technological capacities we have.

Similar to Tom Douglas, to whom we will turn in the next section, Persson and Savulescu also understand moral enhancement as our *motivation* to act morally (Ibid., 167). They forward the steady decrease in racism through our evolution as an example of such a motivationally determined understanding of moral enhancement: the role of racial distinction to signify a lack of kinship by marking off strangers from neighbors has been gradually losing its biological significance, enabling us to comprehend the moral falsity of racism (Ibid., 168). Since moral features are not a social construct, but are based in our biological makeup (Ibid., 168), Persson and Savulescu conclude that the potential hazards of cognitive enhancement are to be kept under control by a "vigorous research program on understanding the biological underpinnings of moral behavior". If these hazards can be controlled successfully, effective forms of moral enhancement are our duty and ought to be *mandatory* (Ibid., 174).

Persson and Savulescu (2008) ground the authors' MBE support on the argument that this type of enhancement will lower the risk of ultimate harm. In Persson and Savulescu (2011) the argument of ultimate harm is elaborated in more detail. Ultimate harm can occur as a consequence of various factors, ranging from the use of weapons of mass destruction to catastrophic climatic changes. The underlying problem is that human moral psychology has been adapted to life in small, cohesive

3.1 Moral Bioenhancement—Categorical Support

societies with primeval technology, while it is unprepared for the moral challenges of a technologically advanced global society. Life in traditional society has developed a bias towards the future among humans, disposing them to care primarily about immediately upcoming events that are relevant to them and their close neighbors. Furthermore, humans are still morally unprepared to respond appropriately to the hardships of larger groups. The development of advanced scientific technology appears to have resulted in the need for a radical change of human moral dispositions.

It is therefore essential that the possibilities of moral enhancement by means of genetic and other biomedical techniques be investigated. The misfit between a limited human moral nature and a technologically sophisticated global society ought to be ameliorated by moral enhancement, in order to achieve restraint, promote cooperation, develop respect for equality, as well as other values that are now necessary for the survival of humanity. And it is precisely scientific progress, the cause of this misfit, that might be employed to address it—by offering means leading to the enhancement of the morality of our behavior. But that is precisely where the caveat ("the bootstrapping problem") is: human beings, that is, those who are in need of moral enhancement, are the ones who have to make a morally wise use of the techniques of moral enhancement (Ibid., 498).

That is how Persson and Savulescu arrive again at the conception of making MBE compulsory. In Persson and Savulescu (2008) they advocated compulsory MBE openly: if hazards with the potential of causing ultimate harm can be controlled successfully, "effective forms of moral enhancement are our duty and ought to be mandatory" (Persson and Savulescu 2008: 174). In their later writings (e.g., Persson and Savulescu 2012), they didn't insist anymore on making MBE mandatory, although from much of what they argue compulsory MBE is being implied. For example, the implication of the above mentioned "bootstrapping problem" is either to abort MBE, to continue to advocate compulsory MBE, or to give arguments in favor of VMBE that might circumvent the "bootstrapping problem". As Persson and Savulescu have not aborted the idea of MBE, nor have they given any reasons favoring VMBE, we can conclude from this that they still are in support of making MBE mandatory.

Persson and Savulescu (2008) can also be criticized from the following perspective—one taken by John Harris and Elisabeth Fenton. If moral enhancement is to take place at a biological level, non-traditional cognitive enhancement is required. Consequently, if we do not continue scientific research into enhancement, we have no hope of achieving the great moral progress that will ensure humans lowering the likelihood of ultimate harm. However, the argument goes, the logic in Persson and Savulescu (2008) appears to lead to an "obstinate predicament": "Scientific progress is both the means of our salvation, as well as the means of our downfall" (Fenton 2011: 148).

In line Persson and Savulescu's notion of compulsory MBE, the "god machine" is imagined as a mechanism that is designed to *impose* morally laudable behavior. Hence, it is entirely in line with a program of compulsory MBE. It is left to every individual to decide for herself whether she wishes to be connected to this device. In that regard, it might appear to be respectful of our free will. But such an impression deceives us. Unlike medication for MBE that we decide to take and can equally decide to stop taking (unless we get addicted to it), the "god machine" hijacks our free will (or that what we believe is our free will) once we get connected to it. This device is charged with policing our thoughts in order to keep us away from acting immorally. Unlike God from the Judeo-Christian and Islamic traditions who keeps our free will intact, the "god machine" resembles more a "police machine" than anything we associate with God from those traditions.[1]

There is also no doubt that the "god machine" would be disinclined to accept our decision to disconnect ourselves from it. Hence, the outcomes of our voluntary decision to take MBE medication and our voluntary decision to connect to the "god machine" are very different. In the first case, our free will remains intact (unless, again, we become addicted to the MBE drug we have been administered), while in the second case our free will is being lost.

Making MBE compulsory in order to lower the likelihood of ultimate harm would deprive humans of their freedom of the will. Depriving humans of this freedom means taking away something that is essential for the existence of humans as moral beings. In actual fact, compulsory MBE, albeit intended to avoid ultimate harm, already inflicts a degree of ultimate harm on humans by depriving them of an essential human quality. *As the "god machine" is an instrument of compulsory MBE, it is a device designed (unintentionally) to inflict harm on humans—if not ultimate, certainly major harm!*

Persson and Savulescu replied to my critique of compulsory MBE (Rakić 2014c) by arguing that freedom is a matter of degree (Persson and Savulescu 2014). Selgelid (2014) found my concept alien to "scalar bioethics". I agree that there are degrees of freedom if we understand freedom as a political concept. We can have more or less free elections, more or less free media. Freedom of the will, however, is a threshold concept. Once limitations are imposed on what we are allowed to will, we cannot call our will free anymore. As soon as an external mechanism decides what we are permitted to will, our freedom of the will has not been limited "to a degree". Our will has ceased to be free.

It is of course possible to argue that a free will does not exist, that it is an illusion. The Libet experiment, as well as later experiments with similar findings, suggest that our decisions might take place before we become aware of them (Libet 1986; Benjamin et al. 1983). More recent findings presented by Lau and colleagues suggest that the perception of intention occurs after executive motor movements (Lau et al. 2007). Wegner reasons along similar lines when discussing how auditory hallucinations produced by schizophrenia seem to suggest a divergence of will and behaviour

---

[1] Soon after the Center for the Study of Bioethics has been founded, it organized in May 2013 the conference "Enhancement: Cognitive, Moral and Mood". The co-organizer was the Oxford Centre for Neuroethics. The event took place in Belgrade and Julian Savulescu and I coordinated the organization. My idea was to open the conference with a debate on enhancement in which Julian Savulescu and John Harris would confront their stances, while Peter Singer would be a discussant of their presentations. This idea was realized and the audience enjoyed a lively debate. It was continued in front of TV cameras in the evening, with Julian, John and myself as participants. During this debate I remember using for the first time the term "police machine" instead of "god machine", suggesting its substantially different role than God has in Judeo-Christian and Islamic traditions.

3.1 Moral Bioenhancement—Categorical Support

(Wegner 2003). Kühn and Brass argue that we might even be unable to veto or halt a decision we have made unconsciously, as this veto might also have taken place on an unconscious or subconscious level (Kühn and Brass 2009). It should be noted, however, that those are internal limitations to our free will. Furthermore, they are limitations that we are not aware of in our direct experience. Compulsory MBE, on the other hand, involves an external mechanism that is designed to affect our will, a mechanism we are conscious of.

The implication is that compulsory MBE, an enterprise that affects what we perceive as our free will, also affects what we perceive as our human identity (which includes us having a free will). Hence, compulsory MBE, affecting what we perceive as our free will, runs contrary to our notion of who we are. *In that sense, it inflicts another essential harm as well: harm, possibly ultimate harm, on our identity as human beings.*

It is of course possible to redefine this identity. However, if we do so by denying the reality of our experience of free will, the redefinition of our human identity, even if possible, would be both difficult and highly traumatic. Hence, it is something that is to be avoided.

Moreover, there is evidence suggesting that the belief that we have in a free will is a significant motivation for us to act morally. Various empirical findings substantiate this. Baumeister and colleagues point to findings that a disbelief in free will decreases helpfulness and increases aggression (Baumeister et al. 2011). Elsewhere, Baumeister et al. argue that trust in free will has behavioral consequences, including increases in socially and culturally desirable acts (Baumeister et al. 2009). In one publication Rigoni et al. show that the readiness potential for acting is lowered in individuals induced to be skeptical about a free will (Rigoni et al. 2012a). In another article Rigoni and colleagues demonstrate that undermining free will can degrade self-control and that it leads to other antisocial tendencies (Rigoni et al. 2012b). Vohs and Schooler provide evidence that mistrust in free will increases the tendency to cheat (Vohs and Schooler 2008). All these findings further strengthen the argument that even the illusion that our will is free should not be easily abandoned. If we abandon it, we might be less prone to act morally. If we believe that freedom of the will is a matter of degree, that we do not fully possess what we have always experienced as a free will, we will be less likely to even try to act morally, achieving exactly the opposite of our goal of moral enhancement.[2]

Furthermore, as the "god machine" deletes our thoughts that it considers as highly immoral, it not only limits our freedom of the will, but also our freedom of thought. It is therefore no wonder that in *Unfit for the Future* (2012) Persson and Savulescu put forward their reservations toward liberal democracy. Indeed, compulsory moral bioenhancement requires a degree of authoritarianism. The "god machine" cannot function in a liberal social setting.

It is not clear who is to decide what qualifies as a "grossly immoral" thought. Let us assume that it is the "moral elite" in a society. Or just a few people who know best

---

[2] The mentioned findings have also been critically assessed by some authors. A discussion of these critiques is beyond this chapter's scope.

where the line between immoral and "grossly immoral" is to be drawn. How can we know who they are? What is the moral elite? Moreover, why should we believe that this elite or just a few of the morally most proper people (whoever they possibly can be) would have the power to be in charge of the "god machine"? Or, for that matter, to be in charge of any type of compulsory MBE program? It is far from certain that the moral and political & financial elite will be congruent. They are likely to be different people. In that sense, the "god machine" cannot be brought into practice. If it were ever developed, it would be a device under the control of the most powerful groups in a society. The same holds for any other compulsory MBE program: it would be run by the most influential social groups, which are by no means necessarily the "most moral" social groups.

Last but not least, compulsory MBE brings into question the conception of love that is being advocated in this book. If love is a matter of our will, as has been argued, compulsory MBE, infringing on our will, would also infringe on our (capacity to) love. If a "god machine" decides which types of love are acceptable and which are "grossly immoral", it decides who deserves our love and who not. In that case it is not we who love. The "god machine" loves instead of us. It loves in our name.

It has been argued already why freedom and love are essential components of morality. *As compulsory MBE diminishes both our freedom and our full capacity to love, also bringing into question our human identity and moral reflection, it is highly detrimental to morality. It achieves exactly the opposite of moral enhancement: moral decline.*

***

Other schools of thought that generally offer categorical support to almost all forms of enhancement, including moral enhancement, are those going under the labels of transhumanism or posthumanism (including techno-progressivism or techno-utopianism). They link the possibility of MBE to "transformative technologies" of the present-day and near future. It is beyond my intentions to discuss them, as such a discussion would violate the thematic unity of this book. For those who are interested in those schools, I strongly recommend an essential reading in the vast literature on the topic: the introductory book by Hughes (2004): *Citizen Cyborg: Why Democratic Societies Must Respond to the Redesigned Human of the Future.*

3.2 Moral Bioenhancement—Hypothetical Support

In this section the stances of some of the most influential hypothetical supporters of MBE will be discussed in more detail. The focus will be on the positions of Tom Douglas and David deGrazia.

Tom Douglas

A step in the direction of bridging the comprehension-motivation gap would be to concentrate on motives for moral behavior. In line with that, Douglas (2008) defines moral enhancement as follows: "A person morally enhances herself if she alters herself in a way that may reasonably be expected to result in her having morally better future motives, taken in sum, than she would otherwise have had" (Ibid., 229). He argues that direct modulation of emotions is something that ought to be pursued in order to morally enhance humans. Examples of moral enhancement Douglas has in mind include a reduction of dislike of certain racial groups, as well a lessening of impulsive violent aggression (Ibid., 231). As an enhancement of morally relevant motivations can have a positive impact on behavior (e.g., less biased behavior towards other races or ethnic groups, less violently aggressive behavior), Douglas' position goes in the direction of an understanding of how to improve not only moral comprehension, but also the morality of behavior. In that sense, he appears to move, purposefully or not, toward the conception of surpassing the comprehension-motivation gap.

In correspondence with his concentration on motives for moral behavior, Douglas also argues against the claim of opponents of biomedical enhancements that such enhancements benefit the enhanced, but harm others. MBE, namely, is an exception to other enhancements, as the morally enhanced person is left with morally better motives than she had previously. Such motives benefit the unenhanced as well. Consequently, MBE is also beneficial to the unenhanced.

Douglas contends that the need to enhance oneself testifies of a motivation not to accept the given, which bio-conservatives tend to consider as a morally sub-optimal motivation. But in the case of moral enhancement, argues Douglas, it is difficult to maintain that we deal with a bad motivation. Much to the contrary, a desire for self-change in the direction of a development of more laudable motives than we previously had would be morally superior to accepting the given.

After having discussed various pros and cons related to moral enhancement, Douglas concludes that the cons do not outweigh the essential pro consisting of moral enhancement leaving us with morally enhanced motives. The reservation Douglas has about MBE that might warrant calling his MBE support "hypothetical" is based on the fact that he does not have a clear preference for compulsory MBE, that is, he does not favour MBE at any cost in order to avoid ultimate harm.

Douglas' conception of moral enhancement that is based on a direct modulation of emotions seems to leave human freedom intact, as it does not limit the quantity of our choices. It limits humans in performing immoral actions, but it does not narrow the spectre of morally acceptable actions. In other words, Douglas thinks that this type of MBE does not reduce the number of choices we have, but prevents us "only" from realizing our wicked options. The question is what this "prevention" entails. How will we be prevented from realizing the wicked options? How much room does Douglas leave for compulsion and how much for freedom? These questions require answers if we want to determine more precisely how qualified Douglas' MBE support is.

Harris (2011) and Verkiel (2017) are inclined to think that Douglas' conception does encroach upon our freedom. But this does not have to be so. As long as we decide ourselves whether or not to undergo MBE, our freedom remains intact. We might have limited our options to perform evil acts, but it is we who have freely decided to do so. That is an essential difference between such an understanding of MBE and compulsory MBE. In the case of compulsory MBE, someone else instead of us decides about whether or not we will be subjected to MBE. Again, a correct understanding of Douglas' treatment of freedom appears to depend on the issue of how unequivocal Douglas is in his opposition to compulsory MBE.

David deGrazia

In deGrazia (2014) the author defends two assertions. First, MBE does not jeopardize our freedom, except in extreme cases. Second, freedom has degrees and it is only one among the values humans hold dear. If we sacrifice some of our freedom in order to get survival, safety, or absence of severe suffering in return, we will opt for lesser freedom. As deGrazia's second assertion implies that even in extreme cases of MBE (such as the "god machine"[3]) a loss of (some?) freedom is justified if we get safety in return, compulsory MBE is justified in such cases. It might then be concluded that deGrazia's position appears to be a justification of the stance of Persson and Savulescu on compulsory MBE. That is however not necessarily so.

I advocated the conception that *only voluntary MBE* is morally acceptable, *except in extreme cases* (Rakić 2014a, b, c). As an extreme case I offered the example of an incarcerated child rapist: his release might be a danger to children, their parents and society overall. As he has forfeited his freedom already (he is spending his life in prison) there is nothing he would lose by being required to undergo MBE (Rakić 2014a: 249). But we can broaden the number of cases in which MBE could be made compulsory. DeGrazia does just that:

Thus, our public policies might support research into and possibly—if and when some forms of MB are demonstrably safe and effective and the state is prepared to make them universally available—encourage <u>or even require</u> the use of certain MBs that help to reduce or eliminate any of the following moral defects:

- *Antisocial personality disorder, a severe failure of motivation*
- *Specific forms of evil such as sadism and intrinsic delight in cheating others, another severe failure of motivation*
- *Lesser forms of moral cynicism that make one more likely than a good person to be corrupted, to cheat on taxes, not to bother to contribute what one agrees is one's fair share, etc.—a more ordinary failure of motivation*
- *Defective empathy as found in persons with narcissistic personality disorder and in others who are very self-absorbed—a failure of insight*
- *Significant prejudice against the interests of those outside one's group of identification, a failure of insight or motivation*
- *An inability to focus on unpleasant realities (e.g., starving children, the abuse of women, the worst conditions of factory farms) that all reasonable people can agree are morally problematic—a failure of insight*

---

[3]In the course of his argument DeGrazia invokes a device very much resembling Savulescu and Persson's "god machine": "Imagine a computer chip that could be implanted in someone's brain such that whenever the agent decided to perform a certain kind of immoral action, he would change his mind" (deGrazia 2014: 366).

3.2 Moral Bioenhancement—Hypothetical Support

- *Weak will or susceptibility to temptation, a failure of motivation*
- *Impulsivity in relation to violence, a failure of motivation*
- *Unwillingness to find common ground when failure to compromise is disadvantageous to all, a failure of motivation*
- *Inability to find creative solutions to difficult problems involving competing interests and values, a failure of insight*
- *Inability to grasp subtle, complicated details that are of undeniable moral relevance (e.g., the ways in which affluent persons benefit economically from the legacies of colonialism and slavery and from current injustices such as treaties with dictators or strongmen who disserve their country-people), a failure of insight* (deGrazia 2014: 364).

This is a rather extensive number of cases to which the need for MBE applies, according to deGrazia. Moreover, deGrazia believes that in these cases MBE may be even required, that is, made compulsory. This makes him occupy a position in-between the idea of Persson and Savulescu that MBE should become compulsory to all of us, and my position that it is to be made mandatory only in extreme cases (the incarcerated repeated child rapist example). DeGrazia's qualifications of compulsory MBE warrant subsuming his stance under hypothetical MBE support.

The question is how sustainable deGrazia's idea is to extend MBE to the cases he stipulates. It seems to be in some of those cases, but only under the proviso that it is not being made compulsory. DeGrazia leaves however the possibility open to make MBE mandatory in the cases he mentions: if MBE is safe and effective, and if the state is prepared to make it universally available, our public policies might encourage or even require its use in order to reduce or eliminate certain moral defects, says deGrazia. But should the moral defects deGrazia stipulates be reduced or eliminated by making MBE compulsory? Let us look at five moral defects that de Grazia proposes for elimination, possibly by making MBE compulsory:

1. "Defective empathy as found in persons with narcissistic personality disorder and in others who are very self-absorbed".
2. "An inability to focus on unpleasant realities (e.g., starving children, the abuse of women, the worst conditions of factory farms) that all reasonable people can agree are morally problematic".
3. "Unwillingness to find common ground when failure to compromise is disadvantageous to all, a failure of motivation".
4. "Inability to find creative solutions to difficult problems involving competing interests and values, a failure of insight".
5. Inability to grasp subtle, complicated details that are of undeniable moral relevance (e.g., the ways in which affluent persons benefit economically from the legacies of colonialism and slavery and from current injustices such as treaties with dictators or strongmen who disserve their country-people) (deGrazia 2014: 364).

It remains unclear how deGrazia imagines the state identifying people with these traits and subjecting them to compulsory MBE. Moreover, not only that some of these defects are insufficiently identifiable, but they are also formulated in very general terms to be a rationale for MBE. DeGrazia does not convince the reader that an elimination of the moral defects he mentions can successfully be carried out by MBE. Eliminating them by making MBE compulsory appears even additionally unsustainable: some rather subtle moral defects ought to be addressed by MBE, which is difficult already, and on top of that an unidentified moral elite has to decide which defects warrant a degree of coercion to eliminate them, while political decision makers have to make sure how this coercion is to be implemented.

DeGrazia also argues that freedom has degrees and that it is only one of the values we hold dear. This argument has already been discussed in the section on Persson and Savulesscu. The main point was that freedom might be a matter of degree, but only if we treat it as a political concept or as one that is analogous to a political concept. We can have more or less free elections, more or less free media. We can have a ruthless dictatorship in which more freedoms are restricted than in a not fully democratic state where the judiciary is influenced by the executive branch but that leaves many freedoms intact. But when we discuss the issue of compulsory vs. voluntary MBE, we do not deal with anything of that kind. We deal with an external mechanism imposing on us what to will. Such an encroachment upon our will cannot be a matter of degree. If a "god machine" intervenes as soon as it discovers a "grossly immoral thought" in us and disables us to follow through on that thought, our freedom to will is not limited "to some degree". We have then been outright *deprived* of our freedom to will, even of our freedom to think.

3.3 Moral Bioenhancement—Hypothetical Opposition

In this section the stances of some of the most influential MBE skeptics will be addressed. It will be devoted to hypothetical MBE opposition, specifically to the positions of John Harris, Nicholas Agar and Robert Sparrow.

John Harris

> *There is no darkness but ignorance.*
> William Shakespeare

The approach to MBE Tom Douglas proposed has been criticized by John Harris who claimed that the means of MBE are rather ineffective.[4] Moreover, direct modulation of emotions would come at an unacceptable cost to our freedom. In fact, we might end up in modulating emotions in ways that actually lead to moral decline (Harris 2011).[5] John Harris is not only against moral enhancement that is being made mandatory, but he appears to fear that even voluntary MBE based on direct modulation of emotions might be detrimental to our freedom.

---

[4] Harris did refer however to Tom Douglas as to the "grandfather of moral enhancement" (Harris 2011). At that time Douglas was about thirty years old.

[5] For Douglas' reply to John Harris, see Douglas (2013).

3.3 Moral Bioenhancement—Hypothetical Opposition

Harris (2011) asserts not only that moral enhancement must in large part consist of cognitive enhancement (Ibid., 106), but that cognitive enhancement ought *not* to be postponed in anticipation of moral enhancement, that is, up to the point when we are morally enhanced to the degree that we cannot inflict ultimate harm upon ourselves. Much to the contrary, if we delay the development of science we expose ourselves to grave dangers, including the danger of ultimate harm (Ibid., 111).[6] In this regard Harris is directly opposed to the position of Persson and Savulescu.

Much of the mass destruction we have been or will be exposed to, is not attributable to malice and is thus not subject to moral intercession, believes Harris. It is rather the consequence of various types of cognitive failure (prejudices, "idiocy" etc.). The most obvious countermeasure to prejudices Harris believes to be a combination of rationality and education, possibly assisted in the future by various new forms of cognitive enhancement (Ibid., 105). The fact that he favors cognitive bioenhancement as one of the means of moral enhancement is a qualification that warrants treating Harris as a hypothetical and not as a categorical MBE opponent; he does not reject MBE in principle, accepting it as a desired effect of cognitive bio-enhancement.

Harris (2011) summarizes his position on moral enhancement as follows:

> So far from being susceptible to new forms of high tech manipulation, either genetic, chemical, surgical or neurological, the only reliable methods of moral enhancement, either now or for the foreseeable future, are either those that have been in human and animal use for millennia, namely socialization, education and parental supervision or those high tech methods that are general in their application. By that is meant those forms of cognitive enhancement that operate across a wide range of cognitive abilities and do not target specifically 'ethical' capacities. (Harris 2011, 102)

Harris argues against MBE on two main grounds:

1. MBE is perilous to our freedom. Balancing freedom and safety, Persson and Savulesvu, Harris believes, give the latter an advantage to an extent that is inappropriate. Harris thinks that our freedom, including our "freedom to fall", ought to be preserved (Ibid., 111). In his own words: "Without the freedom to fall, good cannot be a choice; and freedom disappears and along with it virtue. There is no virtue in doing what you must" (Ibid., 104).
2. The "village idiot" can also cause (ultimate) harm, even more than malevolent people (Ibid., 108–110). It is therefore not MBE, but cognitive bio-enhancement we must concentrate on.

John Harris *might* be partially correct when ground 2 is concerned. If he is indeed right, the very fact that the village idiot can cause more harm than malevolent people does not mean that MBE is unnecessary. Cognitive enhancement should indeed not be postponed until moral enhancement has reached a satisfactory level. In actual fact, both the village idiot and educated people can pose a danger to humankind: both can cause ultimate harm.

---

[6] Harris does not use the term "ultimate harm" in the article that is referred to here, but part of what he describes in it as the dangers humanity faces, does indeed denote what is understood by Persson and Savulescu as "ultimate harm".

Harris appears to be wrong, however, when arguing that cognitive enhancement is sufficient for moral behavior. Educated people are not necessarily moral people, as we all know. It is even not certain that educated people are on average more moral than uneducated people. A positive correlation between education (or intelligence) and moral behavior is highly doubtful. Harris does not wish to recognize the importance of the discrepancy between what we think is right and how we actually behave. He diminishes the relevance of the comprehension-motivation gap, in spite of the fact that he *occasionally* does recognize it—for example, in one instance Harris writes about this gap along the following lines: "Racism still remains widespread but is almost everywhere deplored and in many countries is also against the law. And of course *it is racist behaviour, not racist beliefs that are the problem, or the main problem*" (Ibid., 105; emphasis added).

Regarding Harris's position that MBE comes at an unacceptable cost to our freedom, it might be replied that as long as we decide ourselves whether or not to undergo MBE, our freedom remains fully intact. In that sense, Harris's criticism of MBE affects only compulsory MBE, not voluntary MBE. Moreover, if humans are being prevented from using the possibility of undergoing MBE, their freedom will be curtailed. The fact that they would be able to decide themselves whether to use means of MBE that will make them unfree, even permanently unfree, should not imply that they must be prevented from having the choice of giving up on their freedom. They have such a choice already. For example, they can decide to bring into power a Grand Inquisitor in the form of a totalitarian, even tyrannical regime. The history of humanity shows that humans have on certain occasions decided to do that. Furthermore, humans have the possibility to inflict various other sorts of harm upon themselves, up to taking away their own lives. Suicide is an option people have and if they commit it successfully, it has a permanent effect. Still, that doesn't imply that they are less free if they can commit suicide and lose their freedom and life forever. On the contrary, they are more free if they have this option. Similarly, voluntary MBE does not infringe upon their freedom, even if they have the option of using it in a way that makes them unfree, even permanently: e.g., selling oneself into slavery is a morally dubious act, but this does not imply that humans should not be given that option (for this issue and various other elaborations Harris developed on freedom and MBE, see also Wiseman 2016).

Nicholas Agar

Agar has frequently been regarded as a philosopher occupying a position between bio-conservatives and bio-liberals. For instance, in *Liberal Eugenics* (2004) he supports reproductive freedom—the right of future parents to pursue enhancement technologies for their prospective children—but only on a voluntary basis. He rejects "radical enhancement".

Agar is critical of our tendency to overestimate the benefits of new (bio-)technologies, arguing that when we imagine a technologically-enhanced future we are inclined to think of much better lives than the ones we currently live. In *The*

3.3 Moral Bioenhancement—Hypothetical Opposition

Sceptical Optimist (Agar 2015a) he contends that such "radically optimistic" imaginative exercises "undersell" the past and "oversell" the future, having a distorting influence on how we think about benefits and risks of (bio-)technological progress.

According to Agar, true human enhancement is not radical enhancement, as "too much enhancement" may lead to a future that will be worse than the present. For example, a future in which we are "too enhanced" may distort our relations with other people—with those who are not radically enhanced. Moreover, radical enhancement brings into question our identities if it leads to "transformative change" (Agar 2014).

Agar offers various examples that support his stance, among else in Agar and McDonald (2017). Someone who grew up in a remote village acquires the possibility to study at a top university that is far away. She might rightly conclude that her studies will alienate her from friends and family who remain in her village: their interests will become diverse and one day she might get a job at a geographical location that will permanently remove her physically from them. Hence, she could have a dilemma whether to study at the far away university or not to follow that path after all.

Another example Agar gives is from the Story of Job (Bible, Old Testament): Job loses his beloved kids and property. After a long time he gets children that are objectively superior to the deceased ones. Will someone in Job's position be happy with such a change if it were offered to him while the children he lost were still alive? They might not have been ideal, but he loved them. Could a person like Job opt for a change of children with whom he had an emotional bond for new children who are physically and cognitively enhanced? Probably not, if he is a man/father with adequate emotions toward his children (Agar and McDonald 2017).

When (moral) enhancement is concerned, the two for us most relevant arguments that stand out in Agar's position are the following ones. First, Agar thinks that "post-persons" (enhanced versions of existing people) are imaginable, but undesirable (Agar 2015b). He uses an argument that he calls "inductive" in order to prove that postpersons might rightfully sacrifice "mere persons" (people like us). Hence, we are to avoid them coming into being. Second, too much moral enhancement can go very wrong. Morality is a delicate balance between empathy and appropriate aggression/rightful retribution. If a moral enhancer were administered to us in a slightly improper dose, the consequences could be dramatic. Administering too much of a *cognitive* enhancer would lead to outcomes that are less dramatic: we only become more cognitively enhanced than we initially planned. With *moral* enhancement it is different. Too much empathy might make us morally worse than we used to be before the moral enhancement (Agar 2013): we could become less capable of understanding the delicate balance between empathy and appropriate aggression—an understanding that is so important for morality. Agar does not categorically oppose MBE, but accepts it only under specific conditions. His MBE opposition is hypothetical.

Wiseman (2016 (*Myth of Moral Brain*)) gives similar arguments. But such arguments have significant weaknesses. The very fact, namely, that in some cases we have a difficulty in balancing empathy and an appropriate level of aggression toward those who act immorally, does not mean that we should not increase our empathy in cases in which we know that we need more empathy in order to act in a morally more apposite way. Difficulties in applying MBE do not imply that we should give up on MBE altogether. Furthermore, if we leave it up to every single typical individual (typical referring here to those who are not incarcerated for certain types of serious offenses) to decide when and how much to morally bioenhance herself, we do something very similar to moral education. We only use the helping hand of biotechnologies (on a non-compulsory basis)—accepting the risk that too much empathy might lead to morally worse outcomes. Accepting such a risk is warranted as we cannot expect to take morally relevant actions that are devoid of any risk of that type. Furthermore, existing MBE technologies are likely to improve in the future and Agar's argument against their (too extensive) use can be expected to become increasingly weaker with the passing of time.

Robert Sparrow

Robert Sparrow is another scholar who is critical of the alleged consequences of MBE, but not of MBE in principle. His MBE opposition is also hypothetical. Sparrow's MBE criticism frequently addresses the social aspects of MBE. One of Sparrow's arguments is that MBE poses a threat to freedom—not for the reasons John Harris gives, but because MBE might lead to the "enhancers" wielding power over the "enhanced" and consequently to an inegalitarian order (Sparrow 2014). In "Egalitarianism and moral bioenhancement" (Sparrow 2014) he argues that a society wide program of bio-technological interventions of the sort required to achieve the purported objective of MBE would necessarily implicate the state in what Sparrow calls a "controversial moral perfectionism." Sparrow's misgiving applies however only to compulsory MBE. If there is no state mandated MBE program, the state would not be implicated in a "controversial moral perfectionism." If the possibility of MBE is something that is left to us to decide freely about, there is no moral perfectionism imposed on us by the state.

We can choose one of the following possibilities:

1. Not to undergo moral bioenhancement at all.
2. To opt for voluntarily moral bioenhancement.
3. To put the state in charge, making MBE compulsory.

Although Sparrow is not *prima facie* against MBE, if he had to select one of the three possibilities, he would be in favor of the first one. But if ultimate harm on the one hand, and MBE on the other, are realistic prospects, it is reasonable to favor either the second or the third option. Sparrow neither argues that ultimate harm is unlikely, nor does he offer cogent arguments showing that MBE is impossible or immoral *in principle*. Hence, we are left with the second and third options: voluntary moral bioenhancement or compulsory moral bioenhancement. Sparrow shows why the third option implicates the state in a controversial manner. But the second option he does not discuss at all, and it is precisely this alternative that can successfully address his concerns (Rakić 2014b). The failure to distinguish in the context of his argument between compulsory and voluntary MBE is a major weakness of Sparrow's lines of reasoning pertaining to MBE.

3.4 Moral Bioenhancement—Categorical Opposition

In this section I will briefly address categorical MBE opposition, by merely announcing the position Harris Wiseman has put forward in a number of his publications, primarily in his book *The Myth of the Moral Brain* (2016). In the chapter that follows (Chap. 4), Wiseman's position will be elaborated and criticized in detail. I will devote extensive attention to his position and my arguments against it, because Wiseman has attempted to offer what might be the most cogent explanation of categorical, *prima facie* opposition to MBE. At the outset it ought to be noted that Wiseman is not a categorical opponent of bioenhancement (such as the bio-conservatives discussed in Chap. 1), but only of *moral* bioenhancement.

Harris Wiseman

The following features characterize Wiseman's position on MBE:

1. Wiseman is highly skdueptical of MBE—not under certain assumptions, but *in principle*.

    a. Wiseman argues that what MBE advocates fail to comprehend is the internal complexity of sophisticated moral virtues humankind currently needs, both in order to enhance its moral evolution and to avoid ultimate harm. This internal complexity pertains to, among else, the conceptions of empathy, altruism, kindness, faithfulness, trust, generosity, wisdom and moral imagination. Wiseman believes that these conceptions depend on which culture or faith tradition one is coming from (see Rakić and Wiseman 2018).
    b. The enhancement of sophisticated moral goods in any general sense is implausible. Trust, for instance, can be a necessary virtue, but we give different types of trust to family members, drivers, banks, states, institutions, spouses etc. Wiseman mentions the Buddhist notion of "basic trust," that has specific meanings which are very different from, for example, the idea of "trust in one's creator" that characterizes Christianity. All these kinds of trust have important qualitative differences. The same is true for any moral good (see Rakić and Wiseman 2018).
    c. The limitations to MBE are not merely technological. This is the reason why Wiseman believes that they cannot be ameliorated in the future. The difficulties with enhancing moral goods are obstinate, as they arise from the internal subtleties of the structure of those goods themselves. This dooms any MBE enterprise *in principle* (see Rakić and Wiseman 2018).

2. Persson and Savulescu, "MBE enthusiasts" Wiseman perceives as some of the most infamous ones, commit various grave mistakes, among which the following one stands out: they try to use the lowering of likelihood of ultimate harm as the grounding rationale for MBE. Conversely, *Wiseman argues that MBE should be grounded on something very different than ultimate harm*. A simple reason is that MBE will not prevent ultimate harm. The influence of the writings of Persson and Savulescu have resulted however in much of the moral enhancement

debate being framed in the context of avoiding ultimate harm, that is, survival. That has sidetracked the whole MBE debate, argues Wiseman.

The next chapter will focus on a more extensive discussion of categorical opposition to MBE, that is, on a more detailed elaboration and critique of the position defended by Harris Wiseman.

References

Agar, Nick. 2004. *Liberal Eugenics: In Defence of Human Enhancement.* Hoboken, NJ: Wiley-Blackwell.
Agar, Nick. 2013. Why is It Possible to Enhance Moral Status and Why Doing So is Wrong? *Journal of Medical Ethics* 39: 67–74.
Agar, Nick. 2014. Truly Human Enhancement: A Pfilosophical Defense Od Limits. *Theoretical Medicine and Bioethics*: 1–4.
Agar, Nick. 2015a. *The Sceptical Optimist: Why Technology Isn't the Answer to Everything.* Oxford: Oxford University Press.
Agar, Nick. 2015b. Moral Bioenhancement and the Utilitarian Catastrophe. *Cambridge Quarterly of Healthcare Ethics* 24 (1): 37–47.
Agar, Nick, and J. McDonald. 2017. Human Enhancement and the Story of Job. *Cambridge Quarterly of Healthcare Ethics* 26 (3): 449–458.
Baumeister, R.F., E.J. Masicampo, and C.N. DeWall. 2009. Prosocial Benefits of Feeling Free: Disbelief in Free Will Increases Aggression and Reduces Helpfulness. *Personality and Social Psychology Bulletin* 35: 260–268.
Baumeister, R.F., A.W. Crescioni, and J.L. Alquist. 2011. Free Will as Advanced Action Control for Human Social Life and Culture. *Neuroethics* 4: 1–11.
Benjamin, Libet, C.A. Gleason, E.W. Wright, and D.K. Pearl. 1983. Time of Conscious Intention to Act in Relation to Onset of Cerebral Activity (Readiness-potential). The Unconscious Initiation of a Freely Voluntary Act. *Brain* 106: 623–642.
deGrazia, David. 2014. Moral Enhancement, Freedom, and What We (Should) Value in Moral Behaviour. *Journal of Medical Ethics* 40 (6): 361–368.
Douglas, Thomas. 2008. Moral Enhancement. *Journal of Applied Philosophy* 25 (3): 228–245.
Douglas, Thomas. 2013. Moral Enhancement Via Direct Emotion Modulation: A Reply to John Harris. *Bioethics* 27 (3): 160–168.
Fenton, Elizabeth. 2011. The Perils of Failing to Enhance: A Response to Persson and Savulescu. *Journal of Medical Ethics* 36: 148–151.
Harris, John. 2011. Moral Enhancement and Freedom. *Bioethics* 25 (2): 102–111.
Hughes, James. 2004. *Citizen Cyborg: Why Democratic Societies Must Respond to the Redesigned Human of the Future.* New York: Basic Books.
Kühn, S., and M. Brass. 2009. Retrospective Construction of the Judgement of Free Choice. *Consciousness and Cognition* 18: 12–21.
Lau, H.C., R.D. Rogers, and R.E. Passingham. 2007. Manipulating the Experienced Onset of Intention after Action Execution. *Journal of Cognitive Neuroscience* 19: 81–90.
Libet, Benjamin. 1986. Unconscious Cerebral Initiative and the Role of Conscious Will in Voluntary Action. *Behavioral and Brain Sciences* 8: 529–566.
Persson, Ingmar, and Julian Savulescu. 2008. The Perils of Cognitive Enhancement and the Urgent Imperative to Enhance the Moral Character of Humanity. *Journal of Applied Philosophy* 25 (3): 162–177.

References

Persson, Ingmar, and Julian Savulescu. 2011. Unfit for the Future? Human Nature, Scientific Progress, and the Need for Moral Enhancement. In *Enhancing Human Capacities*, ed. by Julian Savulescu, Ruud Ter Meulen, and Guy Kahane, 486–500. Oxford: Wiley-Blackwell.

Persson, Ingmar, and Julian Savulescu. 2012. *Unfit for the Future*. Oxford: Oxford University Press.

Persson, Ingmar, and Julian Savulescu. 2014. Should Moral Bioenhancement be Compulsory? Reply to Vojin Rakic. *Journal of Medical Ethics* 40 (4): 251–252.

Rakić, V. 2014a. Voluntary moral enhancement and the survival-at-any-cost bias. *Journal of Medical Ethics* 40 (4): 246–250.

Rakić, V. 2014b. Voluntary Moral Bioenhancement Is a Solution to Sparrow's Concerns. *American Journal of Bioethics* 14 (4): 37–38.

Rakić, V. 2014c. We Can Make Room for SSRIs. *American Journal of Bioethics: Neuroscience* 5 (3): 34–35.

Rakić, V., and Harris Wiseman. 2018. Different Games of Moral Bioenhancement. *Bioethics* 32 (2): 103–110.

Rigoni, D., S. Kühn, G. Sartori, and M. Brass. 2012a. Inducing Disbelief in Free Will Alters Brain Correlates of Preconscious Motor Preparation: The Brain Minds Whether We Believe in Free Will or not. *Psychological Science* 22: 613–618.

Rigoni, D., S. Kühn, G. Gaudino, G. Sartori, and M. Brass. 2012b. Reducing Self-control by Weakening Belief in Free Will. *Consciousness and Cognition* 21: 1482–1490.

Selgelid, Michael J. 2014. Freedom and Moral Enhancement. *Journal of Medical Ethics* 40 (4): 215–216.

Sparrow, Robert. 2014. Egalitarianism and Moral Enhancement. *American Journal of Bioethics* 14 (4): 20–28.

Verkiel, Saskia. 2017. Amoral Enhancement. *Journal of Medical Ethics* 43 (1): 52–55.

Vohs, K.D., and J.W. Schooler. 2008. The Value of Believing in Free Will: Encouraging a Belief in Determinism Increases Cheating. *Psychological Science* 19: 49–54.

Wegner, D.M. 2003. The Mind's Best Trick: How We Experience Conscious Will. *Trends in Cognitive Sciences* 7: 65–69.

Wiseman, Harris. 2016. *The Myth of the Moral Brain*. Cambridge, MA: MIT Press.

Chapter 4
Categorical Opposition to MBE: Harris Wiseman

In this chapter Wiseman's position will be contextualized in light of MBE supporters and those MBE skeptics who have reservations regarding MBE, but who don't go as far as Wiseman to reject MBE *in principle*. Wiseman's *The Myth of the Moral Brain* contains a systematic overview of MBE and a development of various arguments against MBE I disagree with. By pointing to and expanding on the disagreements I have with Wiseman's book and some other of his writings I hope to make the VMBE position additionally plausible.

The Myth of the Moral Brain is a thorough approach to MBE in which Wiseman has succeeded to nuance various issues pertaining to MBE. At the same time, he has himself adopted a radical form of MBE skepticism, rejecting any merit that MBE literature has offered. Wiseman is strongly opposed to the position of Persson and Savulescu, but labels also some nuanced approaches to the possibility of MBE as "MBE enthusiasm".

Before discussing my essential arguments against Wiseman's approach, I will emphasize three fundamental issues we agree on:

1. Moral values are a complex system, empathy and decreased aggression certainly not being the only ones. In some cases, a certain degree of aggression (as a proactive refusal to accept injustices that surround us) is morally appropriate. MBE technologies focus however primarily on empathy and a lowering of aggressive impulses. That is not enough. This sort of reductionism is a significant predicament these technologies face. Although empathy is essential for morality (and if we think that we lack it, it is not clear what can possibly be inappropriate to have at our disposal the help of bio-technologies that we can use in order to bring our actions in line with what we believe is morally right), moral values are far from exhausted by empathy. This argument is being discussed from various angles in different sections of this book.
2. Ultimate harm should not be the grounding rationale for MBE. It is something we should try to avoid, but not at any cost (e.g., not at the cost of diminishing our freedom of the will, our capacity for true love, our human identity, our moral reflection—by allowing compulsory MBE, and thereby already inflicting

a degree of ultimate harm upon ourselves (e.g., Rakić 2014, 2017). MBE might help us in lowering the risk of ultimate harm. My discussion with Persson and Savulescu took place in that context. Persson and Savulescu, however, do treat ultimate harm as the grounding rationale for MBE.

3. The "moral brain" is a myth, at least to the extent that morality cannot be reduced to the functions of that organ (or to molecules, as John Harris formulated it (Harris 2016)). Hence, moral issues are too complex to be dealt with only by medication that enhance our empathy or that decrease our aggressive impulses. Moreover, a stronger motivation to behave more morally is not sufficient for the development of desirable moral traits at the level of our cognition.

Those are the essential *similarities* between Wiseman's position of Categorical MBE Opposition (CMBEO) and the VMBE position. The *differences* between his position and the conception of VMBE can be used however as a groundwork for showing the disparities between two fundamentally dissimilar approaches to MBE.

4.1 Where CMBEO and VMBO Disagree

The following three fundamentally different viewpoints between CMBEO and VMBE are the groundwork of all other disparities between these two positions:

I. Wiseman's CMBEO and the conception of VMBE assign very different weights to the discrepancy between what we do and what we believe we *ought* to do.
II. For CMBEO there is no fundamental difference between compulsory and voluntary MBE.
III. Wiseman insists on the importance of what is currently real in MBE technologies at the expense of what future MBE technologies might offer, while any type of MBE support, including VMBE support, is more oriented to the future.

On the surface of Wiseman's argumentation, the following difficulties can be distinguished:

1. Possibly the most forceful formulation of Wiseman's position is the following:

> Though moral enhancement's functioning would depend on its scaffolding, no moral enhancement can create a moral scaffolding. This is a meta-problem beyond moral enhancement's power and scope, but upon which the value of moral enhancement entirely relies. Moral enhancement might help augment a given vision of the good, but it cannot itself create a vision of the good, and relies on there already being a worthy vision of the good in place to scaffold its use. One would need an already morally laudable scaffolding if the prospects for moral enhancement are to be appropriated in a morally laudable way. (Wiseman 2016: 185)

4.1 Where CMBEO and VMBO Disagree

There is no doubt that an appropriate moral scaffolding is needed in order to know which type of moral behavior to enhance. But since we have a sense of good and evil, we have a moral scaffolding already. Had we not such a moral scaffolding, there wouldn't be a discrepancy between how we behave and how we think we *ought* to behave. We are aware of this predicament; we know about the comprehension-motivation gap; we know about the quandary of the Garden of Eden. *Morality is therefore Janus-faced*: one side of the face of Janus is moral reasoning, the other side of his face is moral behavior. The area in-between the two sides is occupied by motivation. That is one of the areas we have for moral enhancement, including moral bioenhancement.

In other words:

a. The Janus-facedness of morality shows that we have a moral scaffolding already.
b. MBE should, among else, be directed toward making morality non-Janus-faced. The morally scaffolded one side of the face of Janus (our comprehension of what is morally right) shows us the direction.[1]

2. Wiseman argues that "persons of imagination, great speakers and persons of vision" in positions of power (who could then create institutions) would do more than medication (Wiseman 2016: 59).

This argument has two deficiencies:

a. Charismatic people who Wiseman describes have apparently not achieved much until now with their imagination, rhetoric and vision, as humanity has ended up facing the possibility of ultimate or a milder form of harm.
b. The history of humankind shows that charismatic people in charge of the structures of power have frequently been anything but a safe option. When in control of important institutions, state agencies in particular, their imagination, rhetoric and visions can have hazardous outcomes.

3. Wiseman believes that "there can be no perfect solution to dealing with the moral evils of the world and this fact must be put at the very foundation of a meaningful moral enhancement discourse" (Wiseman 2016: 63). Moreover, "there is no solution to the possibility of malevolence-caused ultimate harm, and there is no final solution to the existence of evil, destruction, suffering, and harm" (Wiseman 2016: 63).

A fast eradication of destructive malevolence is indeed an illusion and humanity ought to emancipate itself from the "survival-at-any-cost bias" (Rakić 2014). This does not mean however that there is no solution to the challenge Wiseman brings up. A solution could consist of a gradual, historical development of morality that can be speeded up by new MBE technologies. A thorough elaboration of Pinker's

---

[1] The previous does not imply however that the comprehension-motivation gap can or should be bridged by MBE only. Traditional forms of moral enhancement (moral education in the first place) can still achieve better results than MBE.

thoughts on the historical decline in violence (Pinker 2011) and Doyle's thesis that liberal states have been on the increase in the previous two centuries (Doyle 1983) would fall outside the scope of this chapter, but the case can certainly be made that a decline in violence and an increase in freedoms (and hence of our opportunity to act morally because we *choose* so) testifies of a historical development of morality. MBE as a supplement (not a substitute) to this development might offer some hope for humanity successfully addressing the existence of evil, destruction, suffering, and harm.

4. Wiseman argues that moral enhancement is not going to provide "some magical cure for the global moral evils that plague mankind":

> ...no one has even begun to articulate any even half-plausible means by which technologically based moral enhancement can hope to motivate or sustain this level of complex, hands-on engagement on the grand scale. ...The bootstrapping problem comes back to us again: there are simply too many powerful interests, including those of the general public, which are premised on such global ills never being remedied. (Wiseman 2016: 65, 66)

It is possible, however, to bring into question the existence of a "bootstrapping problem". A morally bioenhanced population might contribute to decision makers behaving more morally in order, for example, to get more votes. In addition to an arguably general historical trend toward more freedom and less violence, the fact that people could voluntarily opt for MBE would contribute to the general public becoming more moral. Furthermore, Wiseman has no reason to think that the effects of individual-scale phenomena on a collective level are entirely unpredictable. Although an understanding of these effects requires us to address collective action problems (such as mass collaboration) there is no doubt that small and incremental enhancements of morality at an individual level are likely to contribute to an enhancement of morality at the collective level as well. A small and incremental enhancement still is an enhancement. Hence, there does not have to be a bootstrapping problem.

Unlike MBE advocates, Wiseman's CMBEO does not expect that MBE technologies will develop. Wiseman's position is that they have reached their zenith already. But why would MBE technologies be an exception to practically all other bio-technologies about which we don't believe that they have reached their peak? The fact that part of the quandary of our existence as moral beings does not reside in our lack of motivation to act morally, but in our comprehension of what is morally right, does not mean that MBE technologies cannot be of help in *motivating* us to *act* morally, or that they cannot offer more in the time to come—if not in the domain of comprehending morality, then at least in the domain of strengthening the motivation of humans to *act* morally.

4.2 The Groundwork of the Disagreements

In this section the mentioned differences between CMBEO and VMBE will be discussed in more depth. At the same time, other objections Wiseman raises against VMBE will be addressed. This elaboration should shed additional light on the groundwork behind the two differing perspectives of CMBEO and VMBE.

I. *Why the face of Janus is a good metaphor and why it is important to bridge the comprehension-motivation gap*

Wiseman argues that comprehension, motivation and behavior are intertwined and that no clear line can be drawn among them. He is therefore disinclined to accept the Janus face and gap/space terminology. Moreover, such a terminology allegedly suggests a spatial context.

We are however dealing with a metaphor here. An essential feature of metaphors is that they should not be taken literally. If they were taken literally, they wouldn't be metaphors. Hence, the face of Janus and the gap/space are not spatial categories. Comprehension, motivation and behavior are indeed processes that are complex, intertwined and not easily separable. But what is the implication of that? It should certainly not be that whenever we encounter a complex system we should satisfy ourselves by stating that it is complex, without analyzing it. Analyzing a complex system frequently means breaking it down into its constitutive components—for the sake of analyzing it properly.

It is understandable that certain scholars have misgivings about an "atomized" approach to the brain in which the functions of certain parts of it are used to explain morality in full. But that does not mean that some of the "atomized" neuroscientific insights into our moral functioning are useless. "Atomization" is common in science and philosophy. In the case of our moral functioning, we also should try to isolate some phenomena (e.g., motivation) and make an attempt to explain them in connection with other phenomena (e.g., cognition, behavior). To insist on the fact that morality is complex and fine-grained is not too revelatory. We need to go beyond that and try to develop a model on the basis of an isolation of certain segments of morality's complexity. What can be found behind this complexity is, metaphorically speaking, the two faces of Janus and motivation as the space in-between the two faces.[2]

One face of Janus looks at morality with a cognitive apparatus. It knows the difference between what ought to be done and what ought not to be done. It consists

---

[2] John Harris uses quite similar wording: "The space between knowing the good and doing the good is a region entirely inhabited by freedom" (Harris 2011: 104). In this context I use the term "motivation" rather than "freedom", because my focus is not only on the freedom to do good and the "freedom to fall", but also on our motivation to act in line with what we consider as morally right—a motivation that leads us to transcend the gap between our notion of morality and our actual behavior. It is however only a difference in emphasis. Perhaps interesting to note is that Harris uses spatial metaphors in an even more emphatic manner than I do ("space", "region"), which should not lead one to believe that Harris had in mind a literally (spatially, geographically) understood meaning of the terms he used.

of a set of moral rules with trans-cultural and trans-historical validity: we know that we should not refrain from helping someone who has helped us, we know that we should not wrong someone who has done only good to us, we know that we should not betray our friends, we know that we should not be cruel to helpless people… This is not something that has much to do with our culture. Perhaps there are not many rules of that sort, but they do exist in our comprehension of the morally right and the morally wrong. Sometimes we may not follow those rules. Our perceived self-interest might, for instance, demotivate us to help a friend who has always been good to us. But we know we ought not to behave in that way. Hence, we will not, for example, brag about having successfully grabbed the only unoccupied seat in a bus, right in front of a woman in the ninth month of pregnancy—because we were tired. We might do that and be ashamed to admit it or we can admit it with a feeling of discomfort, but unless we have a certain type of personality disorder we will not brag about how quick and strong we were in such a situation.

After having acted in a way we believe is not morally right, we can feel bad about it and possibly even talk about having done something morally inappropriate. Or we can push aside our conscience and pretend that we haven't done any wrong (before the outside world or even before ourselves). In the case of pretension another interpretation of the double-faced Janus metaphor comes to mind: hypocrisy.

As there are various examples of cases when we know what is morally right and do not act in accordance with this knowledge, there is room for MBE interventions in the space between the two sides of the face of Janus, that is, in our motivation. Even if the number of such cases were relatively small, the repercussions would be large. Namely, by bridging the comprehension-motivation gap, bio-technological advances can limit moral issues to the cognitive realm: after having brought our actions in line with our comprehension of morality, the only remaining space for moral enhancement would be at the level of our cognition; we would cease to have problems with our motivation (that is, we would act as we believe we should), and consequently we could fully concentrate on our comprehension of morality.

At the point when moral issues can be limited to the cognitive realm, moral enhancement would become identical to cognitive enhancement. More precisely, all moral enhancements would become a subset of cognitive enhancements. Only then would John Harris be right in equalizing moral enhancement with cognitive enhancement. But as long as morality continues to be Janus-faced, that is, before humans supersede the comprehension-motivation gap, any equalization of MBE with cognitive bioenhancement would be wrong.

There is one remaining essential point to be made in this context: not all cognitive challenges that morality makes us face are of the same or even similar types. Some can indeed be dealt with by us divesting ourselves of our prejudices, incompetencies and idiocies. Others are more complicated however. They depend on the type of moral values we adopt (e.g., deontological, utilitarian, religious, those based in virtue ethics). In those cases we might have to make a trade-off between different moral values. In cases in which the Janus face metaphor is invoked, however, we do not have moral dilemmas at the cognitive level. In those cases it is our motivation that is not sufficiently strong to make us act in accordance with what we think is morally right.

4.2 The Groundwork of the Disagreements

It is important to deal with these cases because there is a solution to the challenges they pose: an appropriate strengthening of our motivation, either by moral education or by MBE.

II. *Understanding compulsory versus voluntary MBE*

Existing realities should be transcended if we want to make them more moral. That is where the idea of bridging the comprehension–motivation gap comes in, as well as the strategy of VMBE as a supplemental means of bridging it. This strategy is directed toward a moral emancipation from the present, which can only happen in the future. Wiseman's exclusive orientation toward currently existing realities fails to take this into account.

The very moment we diagnose a misfit between the is and the ought we think about bringing the is in line with the ought, which already is future oriented thinking. The present is to be diagnosed and its moral ailments are to be cured in the future. It is common for moral philosophers to try to envision the future as a morally enhanced present. The inability to think beyond the present is misguided ethics.

In the conception of VMBE, the ones who are most likely to surpass the comprehension–motivation gap will be those who voluntarily decide to become better and possibly embark for that reason on the path of MBE. The challenges of motivating people to voluntarily opt for MBE are undeniable. Moreover, it is extremely unlikely that an individual who can freely decide how to act will always act in line with what she believes is morally correct. Moral automata connected to a type of "god machine" might do so, but free individuals won't. Hence, the conception of "bridging the gap" is an ideal scenario. But such an ideal scenario, even if not 100% achievable, is a useful one because it gives us pointers to the type of behavior we ought to aspire and the type of MBE interventions we can utilize in order to realize such ideal aspirations.

VMBE does not encroach upon our free will (or upon our illusion of a free will, if we adopt the counter-intuitive thesis of Libet and colleagues that we don't have a free will). The notion of free will is an essential building block of our human identity and even if it is an illusion, it is an illusion we should cherish (see Chap. 3). Moreover, VMBE is an ideal supplement of traditional moral enhancement. Everyone can decide for herself whether she wishes to become a better person and accept the benefits and risks that might go with it. Compulsory MBE cannot offer anything even similar to that. A ban on MBE would also be absurd. Why should we ban something that is just a supplement that we can elect not to accept? Why should MBE as an elective supplement, if *safe*, raise any concerns?

Another unnecessary concern Wiseman has about MBE is that it is a strategy that finds for every individual a place in a pre-designed social mechanism (like a watchmaker finds a place for every screw in the clock he assembles (Rakić and Wiseman 2018)). This concern is founded only under the assumption that MBE is compulsory. In the conception of VMBE, there is no place for every individual like in a clock, precisely because MBE is voluntary and everyone is free to make her own judgment and choice whether to opt for MBE. No social roles are imposed there.

VMBE implies that it is the responsibility of every specific individual to decide how far to go in enhancing certain moral dispositions, including where to draw a line moral perfectionism should not be allowed to pass. Compulsory MBE, on the other hand, aimed at taking full control over individuals who are to be coerced into MBE that should lower the likelihood of ultimate harm, loses free individuals, getting moral automata in return.

VMBE will of course not have rapid revolutionary effects or significantly lower the likelihood of ultimate harm in the short run. It can only bring about *some* small, piecemeal, incremental changes at the level of *some* individuals, that with the passing of time *might* have aggregate effects (effects that are to be understood as non-linear and in a context that is sensitive to collective action problems). Hence, the VMBE approach is in that sense not pretentious: it aspires incremental changes with obvious effects that might be visible in the future. In the meantime we will have to learn to live with the danger of ultimate harm or a milder form of destruction that humans can inflict upon themselves.

Wiseman correctly thinks that no sharp lines of distinction can be drawn between compulsory and voluntary MBE. For example, in some cases there is a compulsory element in what appears to be voluntary. The proposal of affirmative action incentives the state can offer to those who decide to morally bioenhance themselves is no exception (Rakić 2012, 2014), as in that context the voluntary contains an element of persuasion resembling compulsion. But such nuances are no reason not to use ideal types of concepts in order to make sense of the complexities of morality. If we understand voluntary and compulsory MBE, as well as MBE skepticism in that light, it is entirely appropriate to conclude that the conception of VMBE covers the middle ground between compulsory MBE and conservative MBE skepticism.

Wiseman's neglect of the relevance of the difference between compulsory and voluntary MBE might be rooted in an overly relaxed understanding of the notions of freedom and liberalism. In fact, he and Persson & Savulescu might have something in common vis-a-vis the conception of VMBE: freedom and liberalism are for them less important, possibly dispensable, possibly even a hindrance to MBE. But if we accept the findings of the Libet experiment (and later related experiments) that aspire to prove that humans do not have a free will, it might be argued that compulsory MBE would not deprive humans of anything. They would remain as unfree as they were before. The safety of humans, possibly even the prevention of ultimate harm at any cost, would then become the only concern. Hence, Wiseman's lax understanding of the importance of freedom and liberalism might bring him precariously close to the position of Persson and Savulescu, the position against which he has been arguing most vigorously.

III.   *The game of the future is morally more relevant than the game of the present*

This brings us now to the issue of the importance of freedom at the aggregate level, that is, to the question how important liberal societies are for the moral enhancement of humankind. There is no doubt that liberal societies are a hindrance to a program of compulsory MBE. In fact, they are incompatible with a program that diminishes our freedom of the will and our freedom to think. But they are not hindrance to VMBE. The argument that a program of VMBE would not be popular among voters is fallacious for two reasons. First, why would voters be against it, if it is merely a matter of choice? Even if the state provided the citizenry with incentives to undergo MBE, it is not clear why such a program could not be sold to voters if MBE and its accompanying incentives were open to everyone and if proponents of MBE could make the case for their ideas (the latter to be expected precisely in liberal democracies). Second, more MBE incentives implies more morally bioenhanced citizens, which implies more votes for MBE programs. Hence, once we get the MBE ball rolling, its effects will increasingly resemble those of a snowball. To reiterate, such effects will not become visible in the short run.

To make a VMBE program work, it is important not to focus merely on what is currently possible, abstaining from an anticipation of what the future can bring. If we say that there is no gap between what we do and what we believe we ought to do, we are mixing up the is and the ought. We say that what we have around us is our moral maximum. The point of moral enhancement is however to transcend the given and to aspire a morally enhanced future. This future can entail less violence (Pinker), more liberal societies and fewer wars (Doyle), or it can one day be something resembling Kant's Ethical Commonwealth (see Chap. 6). But in all cases it is something that aspires to be morally superior to what we currently have. Consequently, any serious thought about MBE has to include a reflection on the future. The very moment we diagnose a misfit between the is and the ought we think about bringing the is in line with the ought, which already is future oriented thinking. In that sense, the game of the future is morally more relevant than the game of the present. The present is to be diagnosed and its moral ailments are to be cured in the future. It is common for moral philosophers to try to envision the future as a morally enhanced present. That is precisely what Wiseman fails to do in his CMBEO.

Machiavelli argues the following:

> But since it is my intent to write something useful to whoever understands it, it has appeared to me more fitting to go directly to the effectual truth of things than to the imagination of it. And many have imagined republics and principalities that have never been seen or known to exist in truth. For it is far from how one lives to how one should live. That he who lets go of what is done for what should be done learns his ruin rather than his preservation. (Machiavelli 1992)

Machiavelli's perspective is one of an adviser to an absolutist ruler of his time. Clearly, a small city state on the Apennine peninsula that is determined to preserve itself against predatory empires surrounding it, has to employ various cunning tactics. As its government is absolutist, the interest of the state and the interest of the ruler are perceived as identical. But the interest of the state does not have to be anything morally desirable, especially if the means for achieving it are immoral. Machiavelli might be right that "letting go of what is done for what should be done" brings about the ruler's "ruin rather than his preservation", but that does not mean that "what is done" is moral. Hence, Machiavelli also appears to confuse the is and the ought.

Similarly, Wiseman tries to make "MBE enthusiasts" "get real". But why assume that existing political and other "grand scale" decision makers would have a chance to morally enhance us, while VMBE does not have a chance? Why should we assume that MBE has reached its zenith if we agree that in certain moral issues the greatest problem is the comprehension–motivation gap, a gap that MBE can help bridging by enhancing our motivation to bring our actions in line with what we believe is morally right?

When it is being argued that the game of the future is *morally* more relevant than the game of the present the idea is being conveyed that for our moral enhancement it is essential not to satisfy ourselves with how we act, but to act as we believe is morally right. As argued already, not all moral failings are a consequence of humans acting in a different way than they believe is morally right. But some of them are. And it is precisely failings of that type MBE can do something about.

It is not sufficient to make MBE enthusiasts and futurists "face reality". The conception of morality as a mere trade-off between different moral (if not political) values is machiavellian, with rules that VMBE does not abide by in the moral game it is playing. It does not abide by them because it focuses on the segment of morality that can be addressed by bridging the comprehension–motivation gap, and not on morality as a trade-off between moral-political values.

In addition to the discussed shortcomings of CMBEO, as seen through the lens of VMBE, Wiseman's reasoning appears to suffer from one additional defect, a defect that is essential. Although Wiseman positions himself as a categorical MBE opponent, that is, one who opposes MBE in principle, the type of MBE he addresses are current MBE practices. He rightly notes that they are not as effective and as safe as we wish them to be. This does not mean however that they will not become better in the future. In fact, Wiseman is against MBE in principle, but does not address MBE in principle. He finds various shortcomings in current MBE reality, and infers from them that MBE is wrong in principle.

In conclusion, the three fundamentally different perspectives Wiseman's CMBEO and VMBE are taking are the following: (1) the former downplays the moral relevance of closing the comprehension-motivation gap; (2) they disagree about the assessment that VMBE as a supplement to traditional moral enhancements is a strategy that has certain potentials; (3) the former focuses on existing moral realities only, while the latter insists on morality having to be oriented primarily to the future—something that follows from its aspiration to morally enhance the status quo.

The third difference is the most critical one as it explains the differing views regarding (1) and (2): the view on the relevance of the gap between what we do and what we believe we ought to do, and the view on the relevance and appropriateness of VMBE as a strategy that might be helpful in bridging this gap. Current realities should be transcended if we want to make them more moral. (3) That is where the idea of bridging the comprehension-motivation gap comes in (1), as well as the strategy of VMBE as a supplemental means of bridging this gap (2). Consequently, the VMBE perspective comes down to taking the present seriously, but at the same time it is directed toward a moral emancipation from the present realities. And that can only happen in the future.

This brings us to the issue of what the present MBE realities actually are. That is going to be discussed in the chapter that follows. In addition, it will be shown why compulsory MBE is ineffective *in principle*.

References

Doyle, Michael W. 1983. Kant, Liberal Legacies, and Foreign Affairs. *Philosophy & Public Affairs* 12 (3): 205–235.
Harris, John. 2011. Moral Enhancement and Freedom. *Bioethics* 25 (2): 102–111.
Harris, John. 2016. *Molecules and Morality*. Oxford: Oxford University Press.
Machiavelli, Niccolo. 1992. *The Prince*. Dover Publications.
Pinker, Steven. 2011. *The Better Angels of Our Nature*. New York: Viking Books.
Rakić, Vojin. 2012. From Cognitive to Moral Enhancement: A Possible Reconciliation of Religious Outlooks and the Biotechnological Creation of a Better Human. *Journal for the Study of Religions and Ideologies* 11 (31): 113–128.
Rakić, Vojin. 2014. Voluntary Moral Enhancement and the Survival-At-Any-Cost Bias. *Journal of Medical Ethics* 40 (4): 246–250.
Rakić, Vojin. 2017. Compulsory Administration of Oxytocin Does Not Result in Genuine Moral Enhancement. *Medicine, Health Care and Philosophy* 20 (3): 291–297.
Rakić, Vojin, and H. Wiseman. 2018. Different Games of Moral Bioenhancement. *Bioethics* 32 (2): 103–110.
Wiseman, H. 2016. *The Myth of the Moral Brain: The Limits of Moral Enhancement*. Cambridge (MA) and London: The MIT Press.

Chapter 5
Realistic Means of Enhancing Morality and Why Compulsory MBE is Ineffecive

A relatively succinct compilation will be offered now of technologies that can have an effect on MBE (Sect. 5.1). The discussion will also show why making the use of *any* MBE technologies compulsory is ineffective in principle (Sect. 5.2). Section 5.1 will have a neurological inkling, while Sect. 5.2 will have a philosophical foundation. Both sections will be in a single chapter, because the argument from Sect. 5.2 builds on Sect. 5.1: there are a number of techniques that can serve MBE, to various degrees of effectiveness (Sect. 5.1), while those that are ineffective *in principle* are MBE techniques that are being made compulsory (Sect. 5.2).

In Sect. 5.1, the technologies that might enhance existing humans morally will be discussed. The formulation "existing humans" is being used in order to denote the human species as it currently is. I exclude from the discussion in this chapter genetic interventions in the domain of moral enhancement of offspring. This issue will be raised in the Addendum after Chap. 6. Moral enhancement technologies for existing humans will be divided as follows:

1. substances/medication that might morally bioenhance humans;
2. other techniques that can have an effect on the moral bioenhancement of humans.

It will be argued that the currently existing substances and technologies that can have an effect on the morality of human behaviour include oxytocin, serotonin/SSRIs, vasopressin, propranolol, transcranial magnetic stimulation and optogenetics. These technologies are promising, but not yet at a stage of development that warrants truly successful moral enhancement of an individual human, let alone humankind. The primary limiting factors are the moral function and contextual dependency of the dispositions they modulate, as well as their efficacy and safety. Nonetheless, already now MBE can achieve certain objectives.

5.1 Substances/Medication and Technologies That can Morally Bioenhance Human

Substances/Medication That can Morally Bioenhance Humans

Among the substances/drugs that affect morality and that can possibly serve as moral bioenhancement techniques, the primary focus here will be on oxytocin, serotonin/SSRIs[1] and vasopressin. A brief interpretation will be offered of their potential roles as moral enhancers.

Oxytocin

Oxytocin is a neuromodulator that is produced by the hypothalamus. It is crucial for the development of feelings of intimacy and plays a key role in sexual reproduction, facilitating orgasm, child birth, lactation and maternal bonding. As it is essential for social recognition and pair bonding, it is frequently referred to as the "bonding hormone" (Mikolajczak et al. 2010).

Oxytocin stimulates empathy. Much of our moral functioning is grounded in empathy. Not surprisingly, therefore, the potential of oxytocin to serve as a moral enhancer has been widely discussed.

Oxytocin proves to control aggression, while stimulating empathy, trust, romantic attachment, fidelity, more appropriate behavior of autistic patients and addicts, generosity, as well as certain forms of cognition. It has a role in social behavior in many species, including humans.

Oxytocin diminishes concern and augments feelings of pleasure, safety and contentment around the mate. It restrains those regions of the brain that are related to behavioral control and fear. Feelings of satisfaction, tranquility and security around the mate decrease the potential for aggressive behavior. A proper control of outbursts of aggression is commonly associated with morally suitable behavior.

Oxytocin also has an impact on our cognitive functioning that is related to trust and possibly to morality. For example, inhalig oxytocin strengthens the inclination of humans to find faces more trustworthy (Theodoridou et al. 2009).[2] Furthermore, healthy male subjects who have inhaled oxytocin show an improved ability to remember human faces, in particular happy faces (De Oliveira et al. 2007; Gustella et al. 2008).[3]

Some studies correlate high levels of plasma oxytocin with romantic attachment. This might have implications for the morality of our behavior. For instance, if a couple is separated for a long period of time, the fear of betrayal might increase. A lack of physical intimacy can augment this fear. If we believe that we risk being exposed to betrayal, we may respond in kind. As this is likely to hurt our partner, it is

---

[1] Selective Serotonin Re-uptake Inhibitors (SSRIs) are substances that build up the extracellular level of serotonin. They are normally used as antidepressants.

[2] One explanation for this is that oxytocin decreases the worry of social betrayal (Baumgartner et al. 2008).

[3] In addition, these subjects show to have a better capacity to recognize fear (Marsh et al. 2010).

not morally right. Oxytocin can step in there in order to help romantically attached couples by decreasing their anxiety during periods of separation (Marazziti et al. 2006).

Moreover, some studies show that oxytocin might promote faithfulness in monogamous couples. In one such study, inhaled oxytocin apparently caused men who were in monogamous relationships to increase the distance between themselves and an attractive woman during a first encounter by 10–15 cm. Single men did not increase this distance (Scheele et al. 2012).

Oxytocin also plays a role in the prevention and correction of morally inappropriate behavior of certain types of addicts. For instance, it inhibits tolerance to many addictive substances (drugs, alcohol), while it also reduces withdrawal symptoms. Addicts are at an increased risk of behaving in a morally improper way. Oxytocin can lower the likelihood of such behavior (McGregor and Bowen 2012). According to some preliminary findings, not only addicts but also autistic patients may benefit from the administration of oxytocin. These findings suggest that after inhaling oxytocin autistic patients show more suitable social behavior (Andaria et al. 2010).

Oxytocin also appears to affect generosity. Some evidence for this is provided by the Ultimatum Game. Ultimatum Games feature in economic experiments in which two players have to decide how to split up a sum of money. Player A proposes how to divide this sum between her and Player B. The latter can say 'yes' or turn down the proposal. In the case that Player B opts to decline, neither player receives anything. If Player B accepts the proposal, the money is divided according to the proposal (Rakić 2014). In a version of this game experimental subjects were ignorant about the role into which they would be placed. Oxytocin proved to increase generosity by 80% (Zak et al. 2007).[4]

Although the mentioned features of oxytocin appear to be promising for MBE, a number of notes of caution are in order. There is some evidence, namely, that oxytocin promotes ethnocentric behavior, combining trust and empathy towards in-groups with suspicion and rejection of out-groups. Hence, it might in certain cases augment xenophobic and various other types of heterophobic behaviors (De Dreu et al. 2011). In that sense, oxytocin does not always boost empathy in a morally unbiased manner.

Furthermore, it has been argued that oxytocin enhances *all* social emotions. There is some evidence, namely, that inhaled oxytocin might increase envy and delight in

---

[4]The issue of generosity turns out to be rather complex in this context. For instance, apart from donating money, people can also "donate" their time for voluntary humanitarian work. This might frequently be an even more sensitive marker of generosity than donating money is (Brooks 2005). Furthermore, it is still not certain which specific biological mechanism is responsible for generosity and which role altruism plays in expressions of generosity. For example, Zak et al. (2007) found that enhanced generosity is *not* caused by altruism.

Several games also suggest that oxytocin, apart from influencing generosity, also has a role in increasing trust and decreasing anxiety, but only in certain instances. In a chancy investment game, experimental subjects to whom oxytocin was administered exhibited high levels of trust two times as often as those in a control group. Experimental subjects who were told that they were interacting with a computer did not react in the same manner, which implies that oxytocin was not simply affecting risk aversion (Kosfeld et al. 2005).

the misfortune of others ("*Schadenfreude*") (Shamay-Tsoory et al. 2009). Spite/envy is not considered to be a morally attractive disposition. Hence, in addition to heterophobia, envy is another morally undesirable feature that oxytocin might possibly stimulate. And again, it is moral reflection that has to step in there in order to subdue morally undesirable traits that oxytocin might augment.

Finally, there are some relevant exceptions to Zak's findings that relate to the role of oxytocin as a moral enhancer which stimulates generosity. Zak et al. (2007) experimented with the following variation of the trust game. They gave oxytocin to one group of subjects and compared them with a control group who were not administered oxytocin. Levels of trust were measured by the sum of money that was being entrusted to a trustee. They increased in subjects who inhaled oxytocin. It turned out that many more of them were willing to trust the other party with all of their money. Moreover, the administration of oxytocin to the trustees also heightened their willingness to return the money they were trusted with. But there was a caveat: around 5% of the tested subjects were utterly unresponsive to being trusted or to stimulation with oxytocin. They either did not release oxytocin upon being trusted with the other's money, or they did not reciprocate when oxytocin was present. Zak referred to these people as to "unconditional non-reciprocators" or "bastards"—the ones who take the money and keep everything for themselves. Zak found that such people have traits of psychopaths.[5]

Serotonin/SSRIs

Serotonin is a neurotransmitter that is biochemically derived from tryptophan and believed to be associated with sensations of comfort and joy. The role that serotonin can play in prosocial behavior consists of an augmented capacity to control powerful emotional impulses that lead to socially aversive outcomes (Davidson et al. 2000; Krakowski 2003). It subdues belligerence and the proclivity to cause detriment, while it encourages fairness. According to studies performed by Crockett (2009) and Miczek et al. (2007), prosocial and affiliative behaviors are associated with normal or even increased serotonin function, while antisocial and aggressive behaviors are associated with impaired or reduced serotonin function. This suggests that serotonin has the potential to serve as a moral enhancer.

This potential might be functionally related to the role of oxytocin as a possible moral enhancer. Crockett et al. (2010) proposes two potential pharmacological pathways for modulating human behavior: a direct way ("bottom-up") involving neuropeptides such as oxytocin, which promote prosocial behavior (commitment, empathy, generosity) and another, indirect way ("top-down"), involving serotonin, which limits antisocial behavior by increasing the aversion to harm others. If this understanding is correct, the implication is that there is a functional interaction between the two pathways.[6]

---

[5] See http://www.theguardian.com/science/2012/jul/15/interview-dr-love-paul-zak; retrieved on 23 January 2018.

[6] The amygdala and cingulum are important brain regions that are involved in the regulation of emotions and the assessment of "social threat". In addition to that, the serotonergic system is a key

Serotonin also appears to have the potential to impact on our cognitive abilities, including the ones related to the enhancement of memory and learning capabilities. Hence, apart from its capacity to enhance the morality of our behavior, there is (still limited) evidence that serotonin might serve as a cognitive enhancer as well (Cowen and Sherwood 2013). In that sense, if serotonin does have the potential to favorably impact on our cognitive competencies related to moral judgments, it has an additional moral value.

There are however also a number of notes of caution in order in the case of serotonin/SSRIs. The safety and effectiveness of SSRIs is questionable. First, SSRIs might cause a variety of toxicity problems, and arguably make the neurochemical situation of the aggressive user worse than it used to be before treatment. Second, SSRIs can be insufficiently effective, inasmuch as when working improperly their effects might cause the same symptoms as the ones they are sometimes treating (e.g., anxiety, aggression, mania). Third, SSRIs can intoxicate their user to such a degree that their powers of moral judgment and self-control become diminished. The gruesome deeds of mass murderers who were on SSRIs or in SSRI kick-off phases, provide disturbing evidence for this. On top of that, some scholars claim, even when working properly, SSRIs only seem to have the desired effects by creating various states of "SSRI-induced indifference" (for an elaboration of these and other pitfalls of SSRIs as moral enhancers, see Wiseman 2014: 17).

Vasopressin

Apart from oxytocin, vasopressin is another neuropeptide that increasingly attracts attention as a potential moral enhancer. Vasopressin is released into the brain by neurons of the suprachiasmatic nucleus. It affects temperature and blood pressure, but also aggression.

Vasopressin is a significant mediator of complex social behavior, in which social recognition plays an important role (Caldwell et al. 2008; Raggenbass 2008). Some studies have shown that social familiarity is associated with a number of cortical areas that process multi-modal audio and visual information—e.g., the temporal cortex, temporoparietal and prefrontal cortex (see Frith and Frith 2003; Gobbini and Haxby 2007; Van Overwalle 2009). Walum et al. (2008) argue that brains of male humans use vasopressin as a reward for lasting mate bonding.

A significant contribution to a better perception of the role of vasopressin in social functioning has been given by Zink et al. (2011). Zink's team made use of fMRI in a study that involved 20 healthy volunteers who received either intranasal vasopressin or a placebo. The subjects were asked to recognize images of known or unknown persons. It turned out that, under the influence of vasopressin, previously unknown faces were faster categorized as "familiar faces".

---

mediator of the neural function that modulates fear and anxiety. Furthermore, projections of the ventromedial prefrontal cortex to the amygdala through the anterior cingulate cortex might have a pivotal role. The complex interlay among the mentioned systems, regions and functional networks facilitates insights into the connection between harmful behavior and distress of victims (LeDoux 2000; Kirsch et al. 2005; Blair 2007).

There is no doubt that vasopressin can impact on social recognition, interpersonal interactions and, in general, the evaluation of what surrounds us. It apparently has the potential to impact on morality by stimulating affection and attachment, at the same time favorably influencing our level of life satisfaction. Moreover, it might favorably affect conditions that are marked by social deficits, including autism and anxiety disorders (see Zink et al. 2011).

Other Substances

A number of drugs are already being prescribed specifically for their choice-altering consequences that affect morality. They include the anti-alcohol-abuse drug disulfuram, the weight loss drug orlistat, as well as anti-libidinal agents that might reduce sexual re-offending (Savulescu and Persson 2012). Furthermore, side effects of some antidepressants and antihypertensives include those that are relevant to moral behavior (Terbeck et al. 2012).

The beta-blocker propranolol is a medicine that might affect morality. The amygdala is involved in emotion processing, including the processing of fear. It is believed that racist feelings are induced by the fear center. Propranolol inhibits the amygdala. When it was administered to experimental subjects, they showed to have lower scores in a wide array of psychological tests that were designed to reveal racist attitudes, as compared to subjects in a control group who took a placebo (Terbeck et al. 2012).

Technologies That can Morally Bioenhance Humans

Not only drugs, but also some technologies can influence choices. They include transcranial magnetic stimulation (TMS), optogenetics, transcranial direct current stimulation (TDCS) and deep brain stimulation (DBS). Such technologies can affect behavior, while some of them have the potential to specifically impact on morality.

Transcranial Magnetic Stimulation (TMS)

> *The attempt and not the deed confounds us.*
> William Shakespeare

Transcranial magnetic stimulation is a brain stimulation technique. It can be used both for treatment and enhancement. Its effect on morality is evidenced in Young et al. (2010), which shows that TMS induced disruption of the right temporoparietal junction (RTPJ) reduces the role of beliefs in moral judgments. The RTPJ is an area that is involved in mental state reasoning. When we judge whether an action is morally appropriate or not, we rely on our capacity to infer the mental state of the actor who performs the action. Young et al. tested the hypothesis that the RTPJ is necessary for making moral judgments. The study showed that interfering with activity in the RTPJ disrupts the capacity to use mental states in moral judgment, especially in the case of attempted harms (Young et al. 2010: 6753).

The question is whether TMS to the RTPJ might completely eliminate the attribution of moral significance to beliefs in our judgment of the ethical value of certain actions. Young et al. found that TMS to the RTPJ significantly reduced but did not eliminate the role of beliefs in moral judgment. Participants continued to judge accidental harms as more morally permissible than intentional harms and attempted harms as more forbidden than non-harms. These findings indicate a persistent inclination to attribute moral significance to beliefs in moral judgments.

"No harm, no foul" is a controversial approach to morality that downplays the relevance of intentions in the judgment of the morality of actions. In this context, two questions come up. First, whether TMS (to the RTPJ) makes us less moral or only more utilitarian. The latter might well be the case to a certain degree if "interfering with activity in the RTPJ disrupts the capacity to use mental states in moral judgment, especially in the case of *attempted* harms" (italics added). Second, whether we can develop TMS interventions that can make us *more* moral. For example, can TMS make us less Machiavellian? We don't have such evidence yet.

Optogenetics

Optogenetics is a technique that utilizes light in order to affect neurons. The targeted neurons have previously been genetically sensitized to luminosity. Optogenetics can impact on our morality through its possible effects on mirror neurons. Mirror neurons have an important role in our behavior as they react both when we perform a certain action and when we register the same action being performed by someone else. In that sense, mirror neurons "mirror" actions of others.

Preston and de Waal (2002), Decety (2002) and Gallese and Goldman (1998) argue that mirror neurons are involved in empathy. If they indeed do have a role in empathy, they have a role in morality. Consequently, as optogenetics is a technique that has the potential to stimulate mirror neurons to fire, it may impact on morality.

Transcranial Direct Current Stimulation (TDCS) and Deep Brain Stimulation (DBS)

Transcranial direct current stimulation is a brain stimulation technique that can be used for the treatment of medical conditions, as well as for the enhancement of normal functioning. For example, TDCS can enhance various forms of cognitive performance, including mathematical and language related abilities, memory, attention and coordination. It can also give new insights into the human brain (e.g., Nitsche et al. 2008). Up to now there is no evidence however that TDCS can impact on morality.

Deep brain stimulation is a neurosurgical technique in which electrodes are implanted into certain brain regions. The electrodes produce electric impulses that regulate abnormal impulses. DBS is used for the treatment of various primarily neurological conditions: epilepsy, dystonia, Parkinson's disease, Tourette syndrome, major depression, obsessive-compulsive disorder and chronic pain (Kringelbach et al. 2007). There is no evidence that DBS has an impact on morality.

5.2 Compulsory MBE is Ineffective

In the previous section it has been shown which of the discussed MBE techniques are more effective than others. Those that are made compulsory are however ineffective *in principle*. In this section it will be shown why.

The meaning of "ineffective" is used here as follows: "ineffective" is inadequate to accomplish an intended purpose. There are various types and degrees of (in)effectiveness, but all are characterized by not being fit for purpose.

The following four possible outcomes of moral enhancement endeavours, combining cognition (moral reflection) and behaviour, can be distinguished[7]:

(a) *Unchanged moral reflection, unchanged behaviour*. This outcome stipulates the retention of the status quo. It neither results in humans making better moral judgments, nor in them behaving more morally. There is no moral enhancement taking place at any level. Outcome a) is therefore to be disregarded in the discussion on moral enhancement.

(b) *Unchanged moral reflection, morally enhanced behavior*. Take the case of "paedophile Jack" (Raus et al. 2014):

> Jack is a man with paedophilic urges who is currently incarcerated for having sexually molested a child. Despite a large amount of therapy, Jack fails to see what is wrong with him interacting with children in a sexual way. It is therefore decided to sedate Jack against his will and bring him to a surgery room. Neurosurgeons implant a chip................that will stop Jack from molesting children (Raus et al. 2014: 268).

Two questions arise immediately: do we have the moral right to subject Jack to "moral enhancement" against his will and, as the intervention is taking place against Jack's will, are we dealing here with moral enhancement in the first place?

Outcome (b) is acceptable to the position that Persson and Savulescu advocate. If we are subjected to moral enhancement unknowingly, or even against our will, even to compulsory MBE, its benefits (e.g., safety) might very well outweigh its detriments. According to such reasoning, the common human is not much different from paedophile Jack when it comes to the issue of whether or not to use compulsion in order to achieve MBE. Both paedophile Jack and the "common human" ought to be subjected to compulsory MBE. In Jack's case in order to prevent sexual reoffending, in the "common human's" case in order to prevent ultimate harm.

The following detriments of compulsory MBE have been addressed already:

1. It diminishes freedom/forfeits freedom of the will.
2. It brings human identity into question.
3. It diminishes the human capacity for genuine volitional love.
4. It renders moral reflection practically superfluous (to this issue will be referred in some more detail in this section).

Moreover, compulsory MBE (not of Jack, but of the "common human") is *ineffective* for two reasons: (1) a mandatory administration of MBE drugs/technologies

---

[7]A few paragraphs in this section build on Rakić (2017a, b).

5.2 Compulsory MBE is Ineffective

renders moral reflection practically superfluous[8]; (2) without moral reflection morality loses much of its meaning; consequently, the beneficial outcomes of the use of MBE do not outweigh its drawbacks to the degree that we could speak of effective moral enhancement.

(c) *Enhanced moral reflection, unchanged behaviour.* Unlike outcome (b) that is behavior-oriented, outcome (c) is cognition-oriented. It is however also controversial, as enhanced cognition in the moral realm (enhanced moral reflection) does not imply enhanced moral behavior.

(d) *Enhanced moral reflection, morally enhanced behavior.* This is the best outcome there is, as it includes both enhanced reflection and enhanced behavior in the sphere of morality. An obvious case of (d) is moral education that results in enhanced moral reflection and enhanced moral behavior. Another uncontroversial case is the application of safe biotechnologies resulting in enhanced moral reflection and enhanced moral behavior—provided that their use is voluntary, that is, that we retain the freedom to decide whether we will use them/continue to use them. If we have the freedom to decide whether we will use biotechnologies to morally enhance ourselves, and if we retain our freedom to decide whether we will continue to use them once we have opted for MBE, our freedom will remain fully intact. Hence, there doesn't appear to be anything controversial in safe moral enhancement biotechnologies that are being used on a voluntary basis: freedom, freedom of the will, moral reflection, human identity, capacity for voluntary love—all being kept intact, even not being diminished.

The main beneficial outcome of compulsory MBE (more safety, according to Persson and Savulescu) does *not* trump its detrimental outcome of, among else, our freedom being diminished and hence our moral reflection being rendered practically superfluous. The reason is that MBE does not even come close to guaranteeing its beneficial effects if it is compulsory, that is, if it is not accompanied by appropriate and usable moral reflection.

We are dealing here with the following deduction:

(1) Compulsory MBE renders the role of moral reflection practically superfluous.
(2) With a practically superfluous role of moral reflection we cannot use MBE in an effective manner.
(3) Consequently, MBE does not lead to effective moral enhancement if it is being made compulsory.[9]

---

[8]"Practical" is being used here as an antonym of "theoretical". The meaning of "practically superfluous" comes close to "functionally superfluous". It denotes something that is unusable.

[9]The reasoning can also be stated as follows:
(1) Compulsory MBE renders the role of moral reflection practically superfluous, which implies that options (a) and (b) are being retained (reflecting the same, acting the same; and reflecting the same, acting differently).
(2) As with a practically superfluous role of moral reflection we cannot use MBE as an effective means of moral enhancement, only option (a) remains (reflecting the same, acting the same).

In order for statement (3) to be true, both statement (1) and statement (2) ought to be true. Hence, if we wish to show that we cannot use MBE effectively by making it compulsory, it ought to be demonstrated that (1) compulsory MBE renders the role of moral reflection practically superfluous, and that (2) with a practically superfluous role of moral reflection we cannot use MBE in an effective manner.

Statement (1) is true because we need moral reflection in order to decide what kind of behavior is morally appropriate and, thus, how we *ought* to behave. If we cannot decide to behave how we think we ought to behave because an external mechanism decides about that instead of us, moral reflection becomes *practically* redundant. It might retain theoretical relevance for us, but in the realm of how we actually decide to behave it becomes a surplus. Compelling humans to subject themselves to moral enhancement deprives them of their decision making power in the sphere of the behavior they could opt for if they were free. Hence, compulsory moral enhancement does not only limit the freedom of humans, their capacity to realize volitional love and their capability to perceive their human identity as the identity of free persons, but it also makes moral reflection unusable, that is, practically superfluous. Moral reasoning becomes practically superfluous in the sense that we lose our freedom to behave in line with it. Consequently, statement (1) is true.

Statement (2) asserts that with a practically superfluous role of moral reflection we cannot use MBE in an effective manner. If moral reflection has no practical significance, what remains is MBE. But how to enhance a moral disposition if we don't use cognition/moral reflection? How can we know which disposition to enhance if we don't use moral reflection? MBE would in that case become a random activity. Hence, in order to use MBE in an effective manner, moral reflection is needed. It cannot be rendered practically superfluous. Consequently, statement (2) is true.

Having shown the truthfulness of both statement (1) and statement (2) it can be concluded that statement (3) is also true: Compulsory MBE cannot be an effective means of moral enhancement.

***

In light of the arguments in this chapter, it can be concluded that MBE is not an illusion. Various MBE technologies can be used already with some success. It can reasonably be expected that at least some of them will become more effective with the passing of time, while new ones will be developed. It has been shown that compulsory MBE is ineffective *in principle*. Hence, what remains as a realistic and effective strategy of MBE is VMBE. That is the main subject of the chapter that follows.

---

(3) As option (a) is not a case of moral enhancement, we cannot use MBE as an effective means of moral enhancement if we make its use compulsory.

References

Andaria, E., J.-R. Duhamela, T. Zallab, E. Herbrechtb, M. Leboyerb, and A. Sirigu. 2010. Promoting social behavior with oxytocin in high-functioning autism spectrum disorders. *Proceedings of the National Academy of Sciences of the United States of America* 107 (9): 4389–4394.

Baumgartner, T., Heinrichs M., Vonlanthen A., Fischbacher U. and E. Fehr. 2008. Oxytocin Shapes the Neural Circuitry of Trust and Trust Adaptation in Humans. *Neuron* 58 (4): 639–650.

Brooks, A.C. 2005. Does Social Capital Make You Generous? *Social Science Quarterly* 86 (1): 1–15.

Caldwell, H.K., H.J. Lee, A.H. Macbeth, and W.S. Young. 2008. Vasopressin: Behavioral Roles of an 'Original' Neuropeptide. *Progress in Neurobiology* 84: 1–24.

Cowen, P., and A.C. Sherwood. 2013. The Role of Serotonin in Cognitive Function: Evidence from Recent Studies And Implications for Understanding Depression. *Journal of Psychopharmacology* 27 (7): 575–583.

Crockett, M.J. 2009. The Neurochemistry of Fairness: Clarifying the Link Between Serotonin and Prosocial Behavior. *The Annals of the New York Academy of Sciences* 1167: 76–86.

Crockett, M., et al. 2010. Serotonin Selectively Influences Moral Judgment and Behavior Through Effects on Harm Aversion. *Proceedings NASUSA* 107: 17433–17438.

Davidson, R.J. et al. 2000. Emotion, Plasticity, Context, and Regulation: Perspectives from Affective Neuroscience. *Psychological Bulletin* 126 (6): 890–909.

De Dreu, C.K., L.L. Greer, G.A. Van Kleef, S. Shalvi, and M.J. Handgraaf. 2011. Oxytocin Promotes Human Ethnocentrism. *Proceedings of the National Academy of Sciences of the United States of America* 108 (4): 1262–1266.

De Oliveira, L.F., C. Camboim, F. Diehl, A.R. Consiglio, and J.A. Quillfeldt. 2007. Glucocorticoid-Mediated Effects of Systemic Oxytocin Upon Memory Retrieval. *Neurobiology of Learning and Memory* 87 (1): 67–71.

Decety, J. 2002. Naturaliserl'empathie (Empathy naturalized). *L'Encéphale* 28: 9–20.

Frith, U., and C.D. Frith. 2003 Development and Neurophysiology of Mentalizinm. *Philosophical Transactions of the Royal Society of London. Series B, Biological Sciences* 358: 459–473.

Gallese, V., and A.I. Goldman. 1998. Mirror Neurons and the Simulation Theory. *Trends in Cognitive Sciences* 2 (12): 493–501.

Gobbini, M.I., and J.V. Haxby. 2007. Neural Systems for Recognition of Familiar Faces. *Neuropsychologia* 45: 32–41.

Gustella, A.J., Mitchell, P.B. and F. Matthews. 2008. Oxytocin Enhances the Encoding of Positive Social Memories in Humans. *Biological Psychiatry* 64 (3): 256–258.

Kosfeld, M., Heinrichs, M., Zak, P.J., Fischbacher, U., and E. Fehr. 2005. Oxytocin increases trust in humans. *Nature* 435: 673–676.

Krakowski, M. 2003. Violence and Serotonin: Influence of Impulse Control, Affect Regulation, and Social Functioning. *Journal of Neuropsychiatry and Clinical Neurosciences* 15: 294–305.

Kringelbach, M.L., N. Jenkinson, S.L.F. Owen, and T.Z. Aziz. 2007. Translational Principles of Deep Brain Stimulation. *Nature Reviews Neuroscience* 8 (8): 623–635.

Marazziti, D., B. Dell'Osso, S. Baroni, F. Mungai, M. Catena, P. Rucci, F. Albanese, G. Giannaccini, L. Betti, L. Fabbrini, P. Italiani, A. Del Debbio, A. Lucacchini, and L. Dell'Osso. 2006. A Relationship Between Oxytocin and Anxiety of Romantic Attachment. *Clinical Practice & Epidemiology in Mental Health* 2 (1): 28.

Marsh, A.A., Yu, H.H., Pine, D.S. and R.J.R. Blair. 2010. Oxytocin Improves Specific Recognition of Positive Facial Expressions. *Psychopharmacology* 209: 225–232.

McGregor, I.S and M.T. Bowen. 2012. Breaking the Loop: Oxytocin as a Potential Treatment for Drug Addiction. *Hormones and Behavior* 61 (3): 331–339.

Miczek, K.A., R.M. de Almeida, E.A. Kravitz, E.F. Rissman, S.F. de Boer, and A. Raine. 2007. Neurobiology of Escalated Aggression and Violence. *Journal of Neuroscience* 27: 11803–11806.

Mikolajczak, M., J.J. Gross, A. Lane, O. Corneille, P. de Timary, and O. Luminet. 2010. Oxytocin Makes People Trusting, Not Gullible. *Psychological Science* 21 (8): 1072–1074.

Nitsche, M.A., L.G. Cohen, E.M. Wassermann, A. Priori, N. Lang, A. Antal et al. 2008. Transcranial Direct Current Stimulation: State of the Art. *Brain Stimulation* 1 (3): 206–223.

Preston, S.D. and F.B.M. de Waal. 2002. Empathy: Its Ultimate and Proximate Bases. *Behavioral and Brain Sciences* 25 (1): 1–20.

Raggenbass, M. 2008. Overview of Cellular Electrophysiological Actions of Vasopressin. *European Journal of Pharmacology* 583: 243–254.

Rakić, V. 2014. Voluntary Moral Enhancement and the Survival-At-Any-Cost Bias. *Journal of Medical Ethics* 40 (4): 246–250.

Rakić, V. 2017a. Compulsory Administration of Oxytocin Does Not Result in Genuine Moral Enhancement. *Medicine, Health Care and Philosophy* 20 (3): 291–297.

Rakić, V. 2017b. Enhancements: How and Why to Become Better, How and Why to Become Good. *Cambridge Quarterly of Health Care Ethics* 26 (3): 358–363.

Raus, K., F. Focquaert, M. Schermer, J. Specker, and S. Sterckx. 2014. On Defining Moral Enhancement, a Clarificatory Taxonomy. *Neuroethics* 7 (3): 263–273.

Savulescu, J., and I. Persson. 2012. Moral Enhancement, Freedom, and the God Machine. *The Monist* 95: 399–421.

Scheele, D., N. Striepens, O. Güntürkün, S. Deutschländer, W. Maier, K.M. Kendrick, and R. Hurlemann. 2012. Oxytocin Modulates Social Distance Between Males and Females. *Journal of Neuroscience* 32 (46): 16074–16079.

Shamay-Tsoory, S.G., M. Fischer, J. Dvash, H. Harari, N. Perach-Bloom, and Y. Levkovitz. 2009. Intranasal Administration of Oxytocin Increases Envy and Schadenfreude (Delight in the Misfortunes of Others). *Biological Psychiatry* 66 (9): 864–870.

Terbeck, S., G. Kahane, S. McTavish, J. Savulescu, P. Cowen, and M. Hewstone. 2012. Beta-Adrenergic Blockade Reduces Implicit Negative Racial Bias. *Psychopharmacology (Berl)* 222: 419–424.

Theodoridou, A., A.C. Rowe, I.S. Penton-Voak, and P.J. Rogers. 2009. Oxytocin and Social Perception: Oxytocin Increases Perceived Facial Trustworthiness and Attractiveness. *Hormones and Behavior* 56 (1): 128–132.

Van Overwalle, F. 2009. Social Cognition and the Brain: A Meta-Analysis. *Human Brain Mapping* 30: 829–858.

Walum, H., L. Westberg, S. Henningsson, J.M. Neiderhiser, D. Reiss, W. Igl, et al. 2008. Genetic Variation in the Vasopressin Receptor 1a gene (AVPR1A) Associates with Pair-Bonding Behavior in Humans. *Proceedings of the National Academy of Sciences of the United States of America* 105 (37): 14153–14156.

Wiseman, H. 2014. SSRIs as Moral Enhancement Interventions: A Practical Dead End. *American Journal of Bioethics: Neuroscience* 5 (3): 1–10.

Young, L., J. Camprodon, M. Hauser, and A. Pascual-Leone. 2010. Saxe: Disruption of the Right Temporoparietal Junction with Transcranial Magnetic Stimulation Reduces The Role of Beliefs in Moral Judgments. *PNAS* 107: 6753–6758.

Zak, P.J., A.A. Stanton, and S. Ahmadi. 2007. *Oxytocin Increases Generosity in Humans*, ed. Sarah Brosnan. *PLoS One* 2 (11): e1128.

Zink, C.F., L. Kempf, S. Hakimi, C.A. Rainey, J.L. Stein, and J. Meyer-Lindenberg. 2011. Vasopressin Modulates Social Recognition-Related Activity in the Left Temporoparietal Junction in Humans. *Translational Psychiatry* 1 (4): e3.

Chapter 6
Voluntary Moral Bioenhancement and Happiness as Its Grounding Rationale: The Best Option on Offer

> *If men cease to believe that they will one day become gods then they will surely become worms.*
> —Henry Miller

It has been shown that moral reflection is requisite for proper MBE. As compulsory MBE renders moral reflection practically redundant, the option that remains is VMBE. Three issues that are essential for the conception of VMBE will be discussed: freedom (of the will) in additional detail, the epistemic fallacy of the wish that the human species ought to survive at any cost, and the conception of happiness as the grounding rationale for MBE.

6.1 Freedom and Survival

The issue of freedom in various types of MBE.

In considering the issue of freedom in the use of MBE technologies we have three possibilities:

1. Not to make them available.
2. To make them compulsory.
3. To make them available, but to leave it up to every individual to decide for herself whether to be subjected to them. ***This possibility I call VMBE. It can be subsumed under hypothetical MBE support: MBE is acceptable—provided that it is voluntary.***

These possibilities have different variants. If MBE is not being made available, interventions directed to it can be prohibited, or in a softer form, research into MBE might not be funded or it might receive insufficient funding. The first variant is unreasonable, as no reasonable grounds can be found to prohibit making people better in terms of their morality. Only if medical interventions leading to it are controversial in terms of their safety and efficacy, does it make sense not to fund specific medicines that might lead to MBE or not to approve them being sold.

Compulsory MBE can take the form of the state making it compulsory for all or for some people. Instead of the state, companies can make MBE mandatory for all or for some of their employees. The reason why the state can opt for making MBE compulsory to all people is to make them less likely to perform immoral acts. Fewer and less drastic immoral acts can mean less crime, but it can also lower the likelihood of humans inflicting ultimate harm or a milder form of self-destruction upon themselves. If the state would make MBE compulsory only to certain segments of the population, it is likely that serious offenders (especially incarcerated habitual offenders) would become important targets of MBE interventions. Companies might have a motive to morally enhance some or all of their employees, but it is difficult to imagine that MBE will be considered as an option to be preferred to financial and other incentives or disincentives companies can use in order to make their employees less likely to commit immoral acts versus them. Moreover, some companies might even favor a certain degree of immoral behaviour of their employees vis-a-vis competing companies.

VMBE can mean that all accountable adults might purchase MBE medication without a prescription from a physician. It can however also mean that it is voluntary only to the degree that someone who wishes to subject herself to it decides so and that a relevant physician agrees with it, prescribing the intervention.

A momentous difficulty is how to motivate people to voluntarily decide to morally enhance themselves. Why would they think that MBE is something that might benefit them? The state can incentivize MBE. It can offer its morally bioenhanced citizens "advantage of opportunity": e.g., tax reductions, retirement benefits, schooling allowances for their children (Rakić 2014). As noted already before, the difficulty with a state incentivized program of MBE is that there are no grounds to expect from morally unenhanced political decision makers to take morally astute decisions. The same argument applies to companies, with the additional problem that incentivizing MBE is unlikely to be a priority of almost any company.

Are there any other incentives for people to decide to morally bioenhance themselves? If there are not, the conception of VMBE is in grave trouble—not as grave as a program of compulsory MBE that is detrimental to our existence as free human beings (among else), but still very serious as the number of people deciding to morally bioenhance themselves will be limited if there are no true incentives that can be offered to them in order to opt for the MBE enterprise. On the other hand, if we have such incentives, that is, if we believe that we have reasons to decide to morally bioenhance ourselves, the conception of VMBE is to be considered as clearly superior to its alternatives of compulsory MBE or no MBE at all. In Sect. 6.2 it will be shown that we do have strong reasons to voluntarily decide to morally bioenhance ourselves. Before that, in this Section a number of other essential aspects of VMBE will be considered.

The three possibilities for the use of MBE technologies can be sub-divided as follows:

6.1 Freedom and Survival

1. Not to make MBE technologies available by:

    (a) Prohibiting MBE.
    (b) Not funding or insufficiently funding research into MBE.

2. To make the use of MBE technologies compulsory by:

    (a) The state making MBE compulsory to all citizens.
    (b) The state making MBE compulsory to some citizens.
    (c) Companies making MBE compulsory to all of its employees.
    (d) Companies making MBE compulsory to some of its employees.

3. To make MBE technologies a matter of citizens deciding themselves about their use by:

    (a) Selling MBE medication over the counter.
    (b) Making MBE interventions/drugs available to those who wish to utilize them—but only in case that a relevant physician has agreed to prescribing them.

1 It has already been shown in this book that a prohibition of moral betterment is absurd if MBE technologies are sufficiently safe. If MBE can be both safe and effective, there is also no reason not to spend resources on research into it.

The argument that MBE in general deprives us of our freedom is fallacious. If we are the ones who voluntarily decide whether to utilize MBE technologies, our freedom remains intact. That is essential. Hence, those MBE technologies that are safe and effective should be made available and resources should be spent on research into a further development of such technologies.

2. Compulsory MBE deprives us of our freedom to decide about how moral we will be. Freedom as a political concept is a matter of degree. We can have more or less free elections or more or less free media. But if an external mechanism decides about how moral we will be, it cannot be asserted that our freedom has been curtailed to a degree. Not being able to will something because an external mechanism disables us to do that does not deprive us of our freedom to some degree. It deprives us of our freedom to will something. No matter how limited this deprivation is, we cannot be considered anymore as free if we are subjected to such type of control. Our freedom to will something is not a scalar concept. It is a threshold concept.

Freedom as a political notion or as a notion that is analogous to a political notion is a matter of degree. Freedom to will is however something that we can have only in full or, if an external mechanism exerts control over it—no matter how limited its interventions are—, not to have it at all. Depriving humans of their freedom implies depriving them of an essential component of their moral existence as human beings. Hence, compulsory MBE for all citizens in a state or for all employees in a company is detrimental to their specifically human existence. In actual fact, Persson and Savulescu's idea to lower the likelihood of ultimate harm by making MBE compulsory makes ultimate harm a reality. Hoping to avoid ultimate harm by compulsory MBE, Persson and Savulescu already inflict severe harm on the existence of humans as humans.

Perssson and Savulescu also argue that MBE will not encroach upon our freedom, because we:

- either lack a completely free will and MBE will thus not make us lose our freedom;
- or we have a completely free will that limits the effectiveness of MBE (Rakić 2014).

But they do not take into account the possibility that we can have an entirely free will that does not limit the effectiveness of MBE. As a matter of fact, we can be morally enhanced in an effective manner without losing our freedom. The reason why this is possible is that our free judgment will always remain the adjudicator of the morality of our actions - even if it is has been effectively subjected to MBE. We are free to decide whether we wish to be morally bio-enhanced. If we wish so, we do not give up our freedom. We only use our freedom to decide to be morally bio-enhanced. Our motives might change if we undergo effective MBE (as do our motives change for a variety of other reasons), but our freedom of the will is not going to be curtailed by it. In other words, VMBE, even if brought about in an effective manner by medication, induces us to act more morally, while leaving our free will untouched[1] (Rakić 2014).

It might be morally justified to make MBE compulsory only to those segments of the population who are a danger to society and who have been left without their freedom already. The example we gave referred to incarcerated habitual offenders who have surrendered their freedom already. Consequently, the state has a moral right to treat them as unfree, that is, to *impose* MBE on them.

It is difficult to come up with a similar example in companies. Making MBE compulsory to certain employees does not appear to be a reasonable option. If some employees behave in ways that the company management deems to be immoral or unacceptable in a different sense, they can be sanctioned by various means. Compulsory MBE can hardly be imagined as a possibility that makes sense in such a context. To make MBE compulsory to all employees faces similar difficulties. Moreover, a company that makes MBE mandatory to all employees, including the most decent ones, would presumably need some explanations to offer regarding the moral and legal feasibility of such a demand. Hence, the only reasonable option that remains among the possible variants of compulsory MBE is (2b): the state imposing MBE on a very limited group of its citizens, that is, on specific types of criminals.

3. It follows from the previous that the vast majority of the population should have MBE interventions at their disposal and that they should be the ones to decide about their use. As these interventions, in order to be as safe and effective as

---

[1] Unless we voluntarily opt for MBE that diminishes our freedom to will—something that has a similar effect on our freedom as making MBE compulsory.

possible, should frequently be supervised by competent medical personnel, (3b) is the variant that in many cases will be the best option for MBE. Still, option (3a) is also acceptable in the case of certain medications. Oxytocin nasal spray is an example of a rather safe substance that might be sold over the counter. It is doubtful that a similar approach is currently warranted in the case of almost all other substances with MBE potential.

All in all, the most reasonable stance on MBE is a combination of (2b), (3a) and (3b): VMBE for almost all citizens, sometimes with and sometimes without a physician's prescription, except in the case of specific types of incarcerated repeated offenders on whom the state has the right to impose MBE.

Survival

> *Only to live, to live and live! Life, whatever it may be!*
> Fyodor Dostoyevski, Crime and Punishment

Ultimate harm prevention at any cost is based on the biological conception that a species (humans in this case) ought to survive at any cost. But that is an unrealistic conception/expectation. We can never fully eliminate the possibility of the (self-)annihilation of humankind. Nuclear, bio-technological and other weapons of mass destruction may end up in the hands of one or more deranged individuals who can inflict ultimate harm with them. A small number of sociopaths is sufficient to cause it. Certain pandemics, either natural or artificially caused, may become uncontrollable. Hence, humans have to learn to live with the idea that severe and even ultimate harms will remain a possibility.

Technological developments cannot and should not be reversed, no matter how intensely we fear that life can be extinguished on our planet, and no matter how much we would like to eliminate that fear. The probability of the annihilation of humankind will never be 0. We can therefore only attempt to keep its likelihood to a minimum (e.g., investing in detecting and monitoring possible pandemics, controlling biological, nuclear or chemical terrorist actions, conducting environmentally responsible policies). But this attempt should fall short of minimizing major harms *at any cost*, including the cost of subjecting the whole population to compulsory MBE (Rakić 2014).

The fact that humans wish to survive has to a large degree to do with their biology. All organisms wish to survive. So do humans. But that does not mean that their survival entails a net balance of goodness. Humankind's annihilation might at some point amount to the annulment of a net balance of badness (e.g., if life on Earth has permanently ceased to be worth living). Then it would be good for humans *not* to survive. At another point, humans might not even have a strong wish anymore to survive. They might even wish not to survive. That can be a stage in their evolutionary development at which they have surpassed the biological need to survive, or to survive at all costs (Rakić 2014).

Survival is not imperative. Sometimes its costs can be too high, both for an individual and for humanity in general. Hence, humanity has to give up on the imperative to survive at any cost. Consequently, even if making MBE compulsory would significantly lower the likelihood of ultimate harm, it is still not a strategy constituting a moral imperative. Hence, we ought to seek a grounding rationale for MBE other than ultimate harm prevention.

It has been argued already that the state can adopt affirmative action policies that apply to those who morally bioenhance themselves. The considerable pitfalls of such a conception have already been addressed. The question is whether there is another motivating factor for humans to subject themselves to MBE. If there is not, the conception of VMBE might face a breakdown—at least as a conception that is superior to all other MBE conceptions.

It will be argued however that there is such a motivating factor. It is happiness that will be proposed as the grounding rationale for MBE. It will be argued that morality and happiness are positively correlated. The implications of this correlation will be discussed. It will be shown that they are of paramount importance for VMBE.

6.2 Happiness: The Reason for Being Good?

So obstinately contradictory is man that you cannot compel him to his advantage, yet he yields before everything that forces him to his hurt.
J. W. Goethe

When I do good I feel good, when I do bad I feel bad, and that's my religion.
Abraham Lincoln

I would believe only in a God that knows how to dance.
Friedrich Nietzsche

Happiness and Goodness

As has been previously discussed, Leo Tolstoy's sentence from *Ana Karenina*, "All happy families are alike; each unhappy family is unhappy in its own way", might be interpreted as implying that the unhappy families would have been happier had they been more moral. In this Section we will see why.

We encountered several times the question what will make humans opt voluntarily for something that they haven't been willing to go for in the past. What will make them decide to use MBE technologies in order to become better, after becoming aware of their inability to become better without the use of MBE as a supplement to traditional forms of moral enhancement? What has changed?

Before turning to these questions, let us try to be more specific about what happiness is. It might be argued that we are happy when we have what we desire.[2] We can desire instrumental and intrinsic goods. Instrumental goods are those goods that are

---

[2]Happiness can denote a short-term affect or a stable feeling. The focus in my argument will be on happiness as a stable feeling. In certain cases it resembles the Ancient Greek notion of *eudaimonia*.

chosen for the sake of another good. An example of an instrumental good is money. It is not desired for its own sake. A person who has been given money but is not allowed to spend it has no reason to be happy to have this money. She might be even unhappy to have been given money she is prohibited to spend. This money has therefore no intrinsic value. We don't desire it for its own sake. Similarly, if we are happy to have money because it gives us a certain status, it is not money that makes us happy but the status we acquire by having money. Again, the value of money is instrumental. What is instrumentally good is desired only if it is believed to help us acquire what is intrinsically good. Conversely, something that is desired for its own sake is intrinsically good. Examples might include the desire to acquire knowledge for its own sake. Or to feel love for its own sake.

The most important question here is, however, whether we might desire goodness for its own sake. Why would we wish to be good for its own sake? An answer that is in line with the above raised arguments about the intrinsic value of knowledge or love would be that goodness in itself makes us feel good. We have therefore reason to desire goodness for its own sake.

Conversely, Persson and Savulescu's position treats goodness as an instrumental value. We have to become better (to be morally enhanced) in order to lower the likelihood of ultimate harm. As the prevention of ultimate harm is essential for our survival as a species Persson and Savulescu are willing to subject all people to compulsory MBE.

Discarding compulsory MBE and arguing solely in favor of VMBE we face the critical question what will motivate humans to voluntarily decide to morally bioenhance themselves. It has already been mentioned several times that a solution based on the introduction of the state as an extrinsic mechanism that would motivate people to subject themselves voluntarily to MBE faces the difficulty that morally unenhanced political decision makers would have to adopt the morally most appropriate policies.

Moreover, a state incentivized MBE program has some elements of coercion in it (Carter 2015). The very fact that those who decide not to undergo MBE are not entitled to the benefits the morally bioenhanced have, might make such a program appear dubious from the point of view of respect of citizens' rights and freedoms. It has to be noted however that such type of incentives stimulate citizens to undergo MBE but do not take away their freedom and right to decide otherwise. They retain a free will to decide whether or not to undergo MBE. The freedom a state incentivized MBE program encroaches upon is political freedom. It does so to a certain degree. But it does not take away our freedom to will something (which a program of compulsory MBE does). It does not compromise freedom (of the will) as a threshold concept.

The conception of a state incentivized VMBE program also treats goodness as an instrumental value if it is based on the idea of avoiding ultimate harm. However, such a program can treat VMBE as an intrinsic good as well—if it aspires goodness not as an instrument for avoiding ultimate harm but as a good that is to be aspired for its own sake. Nonetheless, no matter whether a VMBE program based on state incentives treats goodness and moral enhancement as an instrumental or intrinsic value, the way it aspires to achieve them uses in both cases an external mechanism (the state) in order to arrive at those goals.³

The following four options are possible:

(a) Goodness as an intrinsic value that is not being externally enforced;
(b) Goodness as an intrinsic value that is being externally enforced;
(c) Goodness as an extrinsic value that is not being externally enforced;
(d) Goodness as an extrinsic value that is being externally enforced;

These four options come down to the following:

Option (a): our desire to be good for its own sake and a corresponding willingness to be good, without the state or any other entity having to be employed in order to make us better.

Option (b): our desire to be good for its own sake without a corresponding willingness to be good—the state or any other entity having to be employed in order to make us better.

Option (c): our desire to be good because we believe that we need goodness in order to achieve certain extrinsic goals, and a corresponding willingness to be good, without the state or any other entity having to be employed in order to make us better.

Option (d): our desire to be good because we believe that we need goodness in order to achieve certain extrinsic goals, without a corresponding willingness to be good—the state or any other entity having to be employed in order to make us better.

Option (d) is the one Persson and Savulescu have in mind. We ought to be good (morally enhanced) because it will lower the likelihood of ultimate harm. But we are not sufficiently motivated to morally enhance ourselves. Hence, an external mechanism (the state) ought to make MBE compulsory.

Option (b) treats goodness as an intrinsic value that ought to be externally enforced. Goodness makes us happy, but something outside ourselves ought to enforce our happiness/moral betterment. The reason why we are not good enough can be that we are unaware of the fact that goodness makes us happy or that we suffer from a sort of *akrasia* (weakness of the will). A solution for this might be counselling. In the first case it would not necessarily be psychological but perhaps rather philosophical counselling. In both cases, however, it is very difficult to imagine an external mechanism of that type that coerces us into happiness. Hence, option (b) is more a theoretical possibility than a realistic scenario.

The position of VMBE accepts both options (a) and (c): we can either aspire goodness for its own sake without external mechanisms being needed to coerce us to be good (a) or we can aspire goodness as a means for an important objective, such as lowering the likelihood of ultimate harm, again without external mechanisms being

---

³An "external" mechanism denotes here a mechanism that is external to us in that it is not voluntary. Incentivization of certain behavior is not external to us in that we decide ourselves whether to use the incentives we have. Only coercion I treat as external manipulation or enforcement (although, of course, the distinction between coercion and incentivization is not always a clear-cut one).

needed to coerce us to be good (c). Option c) is our voluntary (even if externally incentivized) decision to opt for MBE in order to avoid ultimate harm. But what is of interest in this chapter is option (a): to desire goodness without having to be externally incentivized/softly coerced into it.

That is precisely where the happiness mechanism comes in. We become aware of the fact that goodness is conducive to our happiness and we have a will that is sufficiently strong that there is no reason for an external mechanism to coerce us into moral enhancement (and consequently happiness). Hence, there are reasons that are not externally enforced why humans may decide to undergo MBE, that is, to opt for VMBE.

It feels good to be good. Goodness has in that sense an intrinsic value. If we feel good because we are good there is no reason to search for extrinsic reasons for morally laudable behavior. There is a relationship between goodness and happiness. This relationship differs from a utilitarian understanding of goodness as a maximization of happiness. We are not good because we want to maximize happiness, as utilitarians would claim. The relationship is rather an inverse one: *we are happy because we are good*.

This is obviously not always the case. For some individuals it is rarely the case. Still, these individuals are a minority. Most of us feel good when we are good. Conversely, being bad tends to make us feel bad. It might even be compared to talking good or bad about people or events that surround us. If we have positive thoughts we tend to feel better than when negative thoughts dominate our mind.

6.2.1 Evidence

Maman used to say that you can always find something to be happy about. In my prison, when the sky turned red and a new day slipped into my cell, I found out that she was right.
Albert Camus

Dunn et al. (2008) argue that spending money on others promotes happiness: "Although personal spending is of necessity likely to exceed prosocial spending for most North Americans, our findings suggest that very minor alterations in spending allocations—as little as $5 in our final study—may be sufficient to produce nontrivial gains in happiness on a given day" (Dunn et al. 2008: 1688). The authors also make an attempt to explain why people don't introduce corresponding changes in their behavior, as well as how to help them in that regard:

Why, then, don't people make these small changes? When we provided descriptions of the four experimental conditions from our final study to a new set of students at the same university (N = 109) and asked them to select the condition that would make them happiest, Fisher's Exact Tests revealed that participants were doubly wrong about the impact of money on happiness; we found that a significant majority thought that personal spending (n = 69) would make them happier than prosocial spending (n = 40) (P < 0.01) and that $20 (n = 94) would make them happier than

$5 (n = 15) (P < 0.0005). Given that people appear to overlook the benefits of prosocial spending, policy interventions that promote prosocial spending—encouraging people to invest income in others rather than in themselves—may be worthwhile in the service of translating increased national happiness[4] (Dunn et al. 2008: 1688).

Kennon Sheldon and Sonja Lyubomirsky argue that we can become happier if we decide so. There are certain volitional or activity changes that can increase our happiness. They include resolving to regularly count one's blessings, pursue meaningful personal goals, or commit random acts of kindness (Sheldon and Lyubomirsky 2004). Elsewhere, Lyubomirski adds the following to this list of happiness stimulating and happiness sustaining activities: making someone else happier, affirming significant values, visualizing a positive future, and savoring positive experiences—in order to durably increase a person's happiness level beyond his or her "set point".[5]

Lyubomirsky endeavours to develop a "science of human happiness". To this end, she focuses on three key questions:

(1) What makes people happy?
(2) Is happiness a good thing?
(3) How can we make humans happier than they are?

For the purposes of this chapter the second question is particularly relevant. Happiness is a good thing because it makes us feel good. In that sense, happiness has an intrinsic value.

But in addition to that, goodness and happiness appear to operate in a circularly supportive fashion. Anik et al. (2009) discusses this relationship in the case of charitable giving. It concludes that giving (generosity) and being kind increases happiness, while happier people are kinder and more generous. Generosity and happiness operate in a positive feedback loop. There is reason to assume that other types of moral behavior (kindness, gratitude, making someone else happy) operate in a similar circularly supportive manner.

Isen and Levine (1972) provide evidence for this: when we feel good we are more inclined to help others. In two studies experimenters induced subjects to feel good by offering them minor pleasures: one group received cookies while studying in a library, whereas members of the other group "incidentally" found a dime in the coin return of a public telephone. It turned out that members of both groups were more helpful than control subjects (Isen and Levine 1972: 384). It also turned out to be the case that subjects became more helpful not only when their good mood was brought about by another person (who handed out cookies), but also when their mood was enhanced in an impersonal manner and by a seemingly accidental event (finding a dime) (Ibid., 386, 387).

Hence, happiness is a good thing for two reasons:

---

[4] Among various other questions, the following also arises in this context. Is the greatest happiness of the greatest number something that makes us happy? If so, we have all reasons to be utilitarians. Utilitarianism would then be in our self-interest. But it is far from certain that utilitarian acting makes us happy.

[5] See http://themythsofhappiness.org/about-the-author/; retrieved on 1 February, 2020.

6.2 Happiness: The Reason for Being Good?

(1) Happiness is intrinsically good because it feels good to be happy.
(2) Happiness tends to stimulate goodness, i.e. morally appropriate behavior.

Related to reason 2: conversely, goodness tends to stimulate happiness. That brings us to an answer to Lyubomirsky's first question, i.e. the question what it is that makes people happy. According to the findings that were discussed in the foregoing paragraphs, it is goodness that tends to make people happy. In line with this, we arrive at an answer to Lyubomirsky's third question, that is, how people can be made happier than they are. This can be achieved by making humans better, by enhancing their morality. In other words, moral enhancement can make humans happier still (Rakić 2015).

6.3 Goodness, Happiness, Self-interest and MBE Incentivization in Synergy

If you want to be happy, be.
Leo Tolstoy

MBE can be our voluntary decision because we have reason to believe that being good enhances our happiness and that it is therefore in our interest. Hence, a specific sort of self-interest might be a grounding rationale for MBE. It has been shown why this self-interest differs substantially from the type of self-interest contained in the desire to lower the likelihood of ultimate harm.

The overall result of individuals deciding to bioenhance themselves transcends specific individuals. The implication of more morally enhanced citizens is an increase in the net balance of goodness in society in general. Additionally, the more people engage in MBE, the likelier it is that society will value goodness. In that case, affirmative action policies favoring the morally bioenhanced would fall on a more fertile soil.

MBE has the potential to bring a mechanism into being that leads not only to moral enhancement that is stimulated by the human need to be happy/feel good, but also to the possibility of states adopting successful policies of MBE incentivization. If an increasing number of citizens become aware of the positive correlation between goodness and happiness and start to value MBE more than they did before, political decision makers will act accordingly and become more inclined to favor policies that incentivize MBE. In such a context, democratic deliberation might bring societies closer to "putting a price on moral enhancement", that is, to appropriately linking certain MBE interventions to corresponding incentives designed to pilot them. State incentivized MBE policies that result from democratic deliberation will also be less prone to be perceived as coercive. In actual fact, our desire to be happy and our understanding that MBE might help us in that has the potential to put into motion a mechanism that can facilitate and legitimize a state incentivized MBE program.

The more information people acquire about the positive correlation between happiness and goodness, that is, the more they become morally educated, the likelier it is that they will voluntarily follow the path of moral enhancement. The more morally enhanced a citizenry is, the better the prospects will be for successful state incentivized MBE policies. The more such policies, the more people will voluntarily opt for MBE. Hence, another circularly supportive mechanism can be established: a circularly supportive moral enhancement mechanism of the citizens of a state and the policies of that state.

A discussion of the reasons why humans do not pursue goodness and happiness to the extent that appears to be in their self-interest would be outside the scope of this chapter. It suffices to be noted that this is the case and that certain existing MBE technologies might be of help to people changing their moral habits, while future MBE technologies might be even more so. The first step is therefore education/awareness raising about the (potential) usefulness of these technologies. The very fact that they are being developed will necessarily intensify discussions about their helpfulness. Awareness about their potentials will be raised. Humans may henceforth become more inclined to use effective and safe moral bioenhancers. Those are the changes MBE technologies might bring about vis-a-vis the past in which they were not available. They may trigger a self-perpetuating mechanism, ultimately leading to the moral betterment of humankind. *This means that happiness can replace the biological principle of survival at any cost/ultimate harm prevention at any cost as the grounding rationale for MBE.*

To reiterate, VMBE does not imply that moral reflection will become less important. MBE technologies are not sufficient for our moral betterment. Nonetheless, when we know what is right, MBE technologies can be of assistance in helping humans act in line with this knowledge. By affecting our motivation to act morally, VMBE has the potential to help humans bridge the comprehension-motivation gap. Successful VMBE can therefore address the predicament of the Garden of Eden.

***

Although ever more efficient and safe MBE technologies are likely to be developed in the time to come, the greatest challenges for moral enhancement that have been addressed in this book were:

- moral reflection cannot be enhanced by MBE technologies, and it is precisely moral reflection that is needed in morally complex contexts;
- as the only true type of MBE (for existing individuals capable of decision making) is VMBE, the question is what will motivate people to voluntarily embark on the MBE enterprise.

The first challenge can be addressed by traditional moral enhancement (moral education). Section 6.2 of this chapter dealt with the second challenge. It has been shown that being good is generally conducive to happiness. Hence, with MBE humans will become more motivated to act morally. Our moral reflection might not improve as much as would be useful, but our moral motivation will obtain a significant boost: we will come ever closer to bridging the comprehension-motivation gap. We will

6.3 Goodness, Happiness, Self-interest and MBE Incentivization in Synergy

gradually return to the Garden of Eden, but this time with knowledge of evil. Knowing evil, however, would not make us act in an evil way. *The reason is that VMBE is likely to bring us ever closer to superseding the comprehension-motivation gap by enhancing our motivation to be good.*

Those humans who opt for MBE will better understand the relationship between morality and happiness. The more they use MBE, the more they will find out that moral behaviour increases their happiness—that it tends to be in their interest. Campaigns against MBE and progress in general slows down not only progress in medicine and public health, but also moral progress and therefore the human pursuit of happiness. If humans don't use effective and safe MBE technologies, their moral development will be retarded. Consequently, they will be less happy than they would have been if their moral development hadn't been held back.

History has shown that humans have not been as good as they could have been. Hence, they have not been as happy as they could have been. In that regard it is warranted to conclude that humanity has behaved in that domain largely irrationally throughout its entire history. In the twentieth century this irrationality has culminated with the development of the capability of its self-annihilation.

Humanity developing the capability to annihilate itself resembles the story of the Garden of Eden in which humans who do not know what evil and death are, learn, by their own will, what these calamities entail. They develop a capacity to become evil, as well as a capacity to die. They decide themselves to be less good than they could have been. Hence, they decide to be less happy than they could have been.

In the frequently cited passage from "Paradise Lost" John Milton's God declares: "............... whose fault? Whose but his own? Ingrate, he had of me all he could have; I made him just and right, sufficient to have stood, though free to fall". As moral aptness and happiness are strongly positively correlated, God's words can be rephrased as follows: "............... whose fault? Whose but his own? Ingrate, he had of me all he could have; I made him just and right, *sufficient to be happy, though free to suffer*".

It is up to humanity to decide whether to use its freedom to be "just, right and happy" (which, as we have seen, are more or less one and the same) or "unjust, wrong and sad", as well as which means to use to achieve the former. If safe and effective MBE technologies can help, humans ought to be given the option to use them—if they wish. Hence, research programs into such technologies ought to be supported. Everything else, ranging from compulsory MBE to forfeiting MBE, are options that I hope to have proven in this book to be very wrong paths to pursue.

This book outlined why it is in the interest of humans to pursue happy, moral, free, reflective and loving lives, as well as how to do that. It has also been shown how VMBE can be of help to humans attaining these goals.

The future will uncover whether humans will act accordingly and, after proper moral education and awareness raising about the benefits of MBE, *decide* to be moral and happy, or possibly face the nightmare scenario of ultimate harm. This scenario is the one *von Trier's* Justine justified on moral grounds. She saw however only one possibility—the worst one. She was unabele to envision the possibility that the "Earth" may become better.

References

Anik, L., Lara B. Aknin, Michael I. Norton, and Elizabeth W. Dunn. 2009. Feeling Good about Giving: The Benefits (and Costs) of Self-Interested Charitable Behavior. Harvard Business School Working Paper.

Carter, Sarah. 2015. Putting a Price on Empathy: Against Incentivising Moral Enhancement. *Journal of Medical Ethics* 41: 825–829.

Dunn, Elizabeth W., Lara B. Aknin, and Michael I. Norton. 2008. Spending Money on Others Promotes Happiness. *Science* 319: 1687–1688.

Kant, Immanuel. 1907. *Religion innerahalb der Grenzen der bloßen Vernunft (Religion Within the Boundaries of Mere Reason)*. Ausgabe der Preußischen Akdemie der Wissenschaften (Ak. 6: 3–202).

Isen, Alice M., and Paula F. Levin. 1972. Effect of Feeling Good on Helping: Cookies and Kindness. *Journal of Personality and Social Psychology* 21 (3): 384–388.

Rakić, Vojin. 2014. Voluntary Moral Enhancement and the Survival-At-Any-Cost Bias. *Journal of Medical Ethics* 40 (4): 246–250.

Rakić, Vojin. 2015. We Must Create Beings with Moral standing Superior to Our Own. *Cambridge Quarterly of Health Care Ethics* 24 (1): 58–65.

Rakić, Vojin. 2018. Incentivized Goodness. *Medicine, Health Care and Philosophy* 21 (3): 303–309.

Sheldon, K.M., and S. Lyubomirsky. 2004. Achieving sustainable new happiness: Prospects, practices, and prescriptions. In *Positive Psychology in Practice*, ed. A. Linley and S. Joseph, 127–145. Hoboken, NJ: Wiley.

ADDENDUM: Combining VMBE and IMBE—A Future Beyond the Garden of Eden

Bababadalgharaghtakamminarronnkonnbronntonne-
rronntuonnthunntrovarrhounawnskawntoohoohoordenenthur-nuk!
–James Joyce
(A sound which represents the symbolic thunderclap associated with the fall of Adam and Eve)

I desire to press in my arms the loveliness which has not yet come into the world.
–James Joyce

Moral enhancement of our unborn offspring is an issue that has not been addressed in any detail in the debate on voluntary and compulsory MBE. It deserves to be mentioned here, at least briefly. This type of enhancement is neither voluntary nor compulsory. Our future offspring does not exist yet. Hence, they are not persons capable of taking morally relevant decisions. Subjecting them to MBE can neither be their voluntary nor compulsory decision. As others decide instead of them, but without compelling them, there is reason to call this type of MBE involuntary MBE (IMBE).

Our offspring cannot have any motivation to be subjected to MBE, but *we* can have such a motivation. If we begin to believe (after having been educated about it) that goodness and happiness are positively correlated, we have reasons to subject our offspring to IMBE and as a result increase the likelihood of them living happier lives than they would have lived without IMBE.

IMBE consists of interventions affecting the unborn. Parents can thus decide whether they would like to genetically engineer their offspring by enhancing their morality. One such possibility is genome editing of the unborn. This can consist of genome editing of the embryo that does not intervene in the germline, but it can also consist of germline genome editing.[1]

---

[1] The focus in this section will be on genome editing, because currently (November, 2020) it seems to be the most promising genetic intervention for moral enhancement. It is likely that more efficient techniques will be developed in the future. Much of what is being asserted in this addendum about genome editing in general, applies by analogy to future genetic interventions aimed at moral enhancement (see also Rakić 2019).

The prospects of genome editing for moral enhancement affecting our empathy (by enhancing it), our violent aggression (by attenuating it), and our moral reflection (by improving it) are possibilities bioethicists should investigate. If such possibilities turn out to be realistic, VMBE is not the only alternative to compulsory MBE. Genome editing for moral enhancement does not subject an individual to compulsion, as there is no individual yet who can take decisions. Parents would be (among) the ones to decide about it. It is also no voluntary act of an individual, for the same reason: because there is not yet an individual who can take decisions.

Why should we focus on empathy, violent aggression and moral reflection? In 2009, in the first gene study of its kind, scientists from the University of Cambridge identified 27 genes associated with Asperger Syndrome and/or autistic traits and/or empathy (Chakrabarti et al. 2009). These findings appeared to be only an indication of later discoveries. In 2017, for example, Warrier et al. provided strong evidence that the ability to detect and understand emotions in others is influenced by our genes (Warrier et al. 2017). Reading, understanding and responding to emotions in others is essential for social interactions; and they are indeed influenced by genetics, as the mentioned study shows. As genes impact on our empathy, it is reasonable to hypothesize that we will be able to develop genetic interventions with the potential to strengthen our empathy. Genome editing is an obvious candidate for such an intervention.

In several prominent murder cases in the United States and Europe, courts have permitted defendants to be tested for the presence of the so-called "warrior gene", and allowed positive results to be considered as a mitigating factor. The gene in question is the neurotransmitter-metabolizing enzyme monoamine oxidase A, abbreviated "MAOA". Abnormalities have been found in brain-imaging scans and in five genes that have been linked to violent behaviour—including the gene encoding the MAOA. Caspi et al. (2002) also found low levels of MAOA expression to be associated with aggressiveness and criminal conduct of young male adults raised in abusive environments. As genes constitute part of the explanation of violent aggression, genome editing of the unborn has the potential to attenuate violent aggression. It is therefore a possible venue for IMBE.

Research conducted at the University of Edinburgh that analyzed genetic data from 20,000 people taking part in a study called "Generation Scotland", has shown that genetic mutations that harm our health may also decrease our intelligence. It appears, namely, that intelligent people have fewer genetic mutations that adversely affect their intelligence and health, rather than more mutations that make them smarter. This implies that being born with a smaller number of bad mutations could be more important to being intelligent than having many mutations that positively affect intelligence. This led scientists to argue in favor of genome editing for cognitive enhancement: editing mutations, they contend, might make people healthier and smarter at the same time.[2] As smarter people have a better capacity for moral reflection than

---

[2] See: https://www.biorxiv.org/content/early/2017/06/05/106203.full.pdf+html; and https://www.newscientist.com/article/2137926-dna-variants-that-are-bad-for-health-may-also-make-you-stupid/?cn=bWVudGlvbg%3D%3D; both accessed on 6 November, 2020.

less intelligent people, cognitive enhancement of the unborn by genome editing has the potential to morally enhance humans (in the realm of understanding morality). It would achieve that goal by IMBE.[3]

As has been argued in this book, the pitfalls of compulsory MBE are that it infringes upon our freedom (of the will), human identity, capacity for volitional love and that it makes moral reflection practically superfluous. Its advantage is that it may lower the likelihood of ultimate harm. IMBE by genome editing does not diminish freedom, it does not affect identity (unborn humans have no consciousness of identity) and it does not infringe upon the capacity for volitional love. Nor does it make moral reflection practically superfluous: as the future genome edited individual will fully retain her freedom, her moral reflection will also be kept intact. If genome editing for moral enhancement engineers morally enhanced individuals, the likelihood of ultimate harm might become lower as well. Hence, IMBE combines the advantages of VMBE (preservation of freedom (of the will), of human identity, of the capacity for volitional love, of uncompromised moral reflection) with the advantage of compulsory MBE (our offspring would be less likely to cause ultimate harm). The domains in which IMBE currently appears to have the potential of success is genome editing designed to increase empathy, to attenuate violent aggression and to improve cognitive functioning, including moral reflection.

A complete program of MBE could consist of a combination of VMBE and IMBE of the unborn. VMBE would be directed toward individuals who have the capacity to comprehend morality. They can decide to morally bioenhance themselves. The positive correlation between morality and happiness might motivate them to do that (Sheldon and Lyubomirsky 2004; Anik et al. 2009; Dunn et al. 2008; Isen and Levin 1972). And the state can offer positive incentives to that effect (see Rakić 2017).

A combined VMBE-IMBE program would offer humans the possibility to embark on MBE, at the same time leaving intact their freedom to decide otherwise. It is to be expected that humans will become increasingly inclined to opt for MBE targeting both themselves and their offspring. In the first case, the more people learn about the positive correlation between morality and happiness, the more people are likely to choose MBE.

In the second case, parents who opt for IMBE of their offspring might have morally enhanced children. These children can reasonably be expected to be happier than they otherwise would have been. Moreover, they can reasonably be expected to be more moral (e.g., more empathetic) in relation to their parents than they otherwise would have been. This might incentivize parents to opt for IMBE. Parents also have a moral right to do so out of respect for the conception of procreative beneficence and because they have a legitimate interest in providing their children with the best options in life, including the option of a (morally) good life.

---

[3]For a more detailed analysis of the potentials of genome editing in strengthening empathy, attenuating violent aggression and improving moral reflection, see Rakić (2019).

References

Anik, L., L.B. Aknin, M.I. Norton, and E.W. Dunn. 2009. *Feeling Good about Giving: The Benefits (and Costs) of Self-interested Charitable Behavior.* Harvard Business School Working Paper.

Caspi, A., J. McClay, T.E. Moffitt, J. Mill, J. Martin, I.W. Craig, A. Taylor, and R. Poulton. 2002. Role of Genotype in the Cycle of Violence in Maltreated Children. *Science* 297: 851–854.

Chakrabarti, B., F. Dudbridge, L. Kent, S. Wheelwright, G. Hill-Cawthorne, C. Allison, S. Banerjee-Basu, S. Baron-Cohen. 2009. Genes Related to Sex-Steroids, Neural Growth and Social-Emotional Behaviour are Associated with Autistic Traits, Empathy and Asperger Syndrome. *Autism Research*, online.

Dunn, E.W., L.B. Aknin, and M.I. Norton. 2008. Spending Money on Others Promotes Happiness. *Science* 319: 1687–1688.

Isen Alice, M., and F. Levin Paula. 1972. Effect of Feeling Good on Helping: Cookies and Kindness. *Journal of Personality and Social Psychology* 21 (3): 384–388.

Rakić, V. 2017. The Issues of Freedom and Happiness in Moral Bioenhancement: Continuing the Debate With a Reply to Harris Wiseman. *Journal of Bioethical Inquiry* 14: 469–474.

Rakić, V. 2019. Genome Editing for Involuntary Moral Enhancement. *Cambridge Quarterly of Healthcare Ethics* 28 (1): 46–54.

Sheldon, K.M., and S. Lyubomirsky. 2004. Achieving Sustainable New Happiness: Prospects, practices, and Prescriptions. In *Positive Psychology in Practice*, ed. A. Linley and S. Joseph, 127–145. Hoboken, NJ: Wiley.

Warrier, V., K.L. Grasby, F. Uzefovsky, R. Toro, P. Smith, B. Chakrabarti, J. Khadake, E. Mawbey-Adamson, N. Litterman, J.J. Hottenga, G. Lubke, D.I. Boomsma, N.G. Martin, P.K. Hatemi, S.E. Medland, D.A. Hinds, T. Bourgeron, and S. Baron-Cohen. 2017. Genome-Wide Meta-Analysis of Cognitive Empathy: Heritability, and Correlates with Sex, Neuropsychiatric Conditions and Cognition. *Molecular Psychiatry* 00: 1–8

The manufacturer's authorised representative in the EU is Springer Nature Customer Service Centre GmbH, Europaplatz 3, 69115 Heidelberg, Germany. If you have any concerns regarding our products, please contact ProductSafety@springernature.com

Printed and bound by CPI Group (UK) Ltd, Croydon, CR0 4YY

25/03/2026

02078197-0011